THE CITY

THE SOCIOLOGY OF URBAN LIVING

SOCIOLOGY OF THE CITY

THE SOCIOLOGY OF URBAN LIVING

HAROLD E. NOTTRIDGE

Routledge
Taylor & Francis Group
LONDON AND NEW YORK

First published in 1972

This edition published in 2007
Routledge
2 Park Square, Milton Park, Abingdon, Oxon, OX14 4RN

Simultaneously published in the USA and Canada by Routledge

711 Third Avenue, New York, NY 10017

Routledge is an imprint of Taylor & Francis Group, an informa business

Transferred to Digital Printing 2007

Transferred to digital print 2013

British Library Cataloguing in Publication Data
A CIP catalogue record for this book
is available from the British Library

The Sociology of Urban Living

ISBN13: 978-0-415-41744-0 hbk (volume)
ISBN13: 978-0-415-41931-4 (subset)
ISBN13: 978-0-415-41318-3 (set)
ISBN13: 978-0-415-86468-8 pbk (volume)

Routledge Library Editions: The City

The Sociology of Urban Living

Harold E. Nottridge

Senior Lecturer in Sociology
Dudley College of Education

LONDON AND BOSTON
ROUTLEDGE & KEGAN PAUL

First published 1972
by Routledge & Kegan Paul Ltd
Broadway House, 68-74 Carter Lane,
London EC4V 5EL and
9 Park Street,
Boston, Mass. 02108, U.S.A.

Printed in Great Britain by
Northumberland Press Limited
Gateshead 8

ISBN 0 7100 7376 3 (c)

ISBN 0 7100 7377 1 (p)

Contents

General editor's introduction

Today sociology is going through a phase of great expansion. Not only is there a widespread general interest in the subject, but there is rapid growth in the numbers of new courses at Universities, Colleges of Further Education, and elsewhere. As a result there is an increasing number of potential readers of introductory textbooks. Some will be motivated by general interest; some will want to find out enough about the subject to see whether they would like to pursue a formal course in it; and others will already be following courses into which an element of sociology has been fused. One approach to these readers is by means of the comprehensive introductory volume giving a general coverage of the field of sociology; another is by means of a series of monographs each providing an introduction to a selected topic. Both these approaches have their advantages and disadvantages. The Library of Sociology adopts the second approach. It will cover a more extensive range of topics than could be dealt with in a single volume; while at the same time each volume will provide a thorough introductory treatment of any one topic. The reader who has little or no knowledge in the field will find within any particular book a foundation upon which to build, and to extend by means of the suggestions for further reading.

It is perhaps difficult to see the continuing and worldwide expansion of urban populations in a way which does not involve a value judgment. Certainly there are many influential voices which condemn the spread of towns and cities as 'urban sprawl' or see in the growth of suburbs with their rows of 'little boxes' a social and cultural desert, just as there are many who see in the city the supporting base for a whole range of cultural activities, and in the new suburbs and housing estates the potential for a growth of 'community' with its implication of a supportive and satisfying network of social relationships. But there are two main reasons why the issues involved need to be examined as dispassionately as possible.

First, the understanding of what is involved in urban living becomes ever more urgent to social scientists and planners alike. There is no reason why sociologists and others should not take up positions which involve their making a value judgment about the desirability or otherwise of various developments in society and their consequences. The essential requirement is only that such standpoints should be based on the best evidence, both theory and data, which is available. As will be seen from this book, such evidence is often hard to come by and the issues are more complex than public controversy would often make them seem. The second reason is arguably even more telling. This is that much controversy in the field of social life arises because of a readiness to concentrate on what may appear to be immediate social problems, and it is because of this that value judgments come so rapidly. On the other hand, a careful study of social structure and social processes, without too much emphasis on whether one is solving a social problem, may eventually lead to a better solution to the problems which arise. From this book the reader will, it is hoped, get a concise idea of what these social processes are in the field of urbanization, and what is the sociologist's approach to them.

A. R. EMERSON

Acknowledgments

The author and publishers would like to thank the following for permission to quote from copyright material:

Constable, *170 Chinese Poems*, Arthur Waley (trs.)

Faber & Faber and Random House Inc., *Collected Poems 1928-1953*, Stephen Spender and *The Age of Anxiety*, W. H. Auden

Jonathan Cape, *The Poetry of Robert Frost*, Edward Connery Lothem (ed.)

Harper & Row, *The Beaneaters*, Gwendolen Brooks

1

What is urban sociology?

The word 'urban' was hardly used in the English language before the nineteenth century. It is briefly defined by the *Shorter Oxford Dictionary* as 'pertaining to town or city life'. It is derived from the Latin 'urbs' a term applied by the Romans to a city—more especially the city of Rome.

The word 'sociology', which is so frequently applied to all kinds of situations, is also of recent origin; the vagueness of its usage in common language and the element of disagreement that occurs among sociologists about the meaning of the word, make it advisable for us at the outset to accept a definition which will be adhered to throughout the whole of this work. We shall consider sociology to be essentially the science of social structure and social relationships. Urban sociology is the sociology of urban living; of people in groups and social relationships in urban social circumstances and situations.

Although the study of people in towns and cities is an important aspect of urban sociology it would be unwise to confine the study solely to towns and cities. It has proved extremely difficult to create a complete sociological definition of the city. One reason for this difficulty lies in the immense variations which exist between towns and cities whether we are looking at existing ones or those known to us in the past. Towns vary in size, density of population, general appearance and the reasons for their development; in their relationship to other places and in a host of other ways. They are often closely linked with the society which has led to their creation. It is not yet possible to find a common urban pattern which will apply, for instance, to a remote Latin American mountain town of some 2,000 inhabitants and equally to the huge industrial complexes of the English Midland conurbation where one town merges into another.

Another question of definition is linked with the problem of boundaries. It is occasionally possible to find historical examples of towns whose walls acted as a limit between the townsfolk and the country people. The townsman in certain German mediaeval towns might say with pride, '*Stadtluft macht frei*', meaning that the townsman was freer than the countryman because he possessed certain rights, but it is doubtful whether we can say that this represented a kind of social distinction which was universally applicable. In many highly urbanized countries today it is hard to demonstrate that there are definite sociological differences between town and country based solely on the location of the inhabitants. One American sociologist has found that in the middle of a large, densely populated industrial city, there were 'villagers' (Gans, 1962). They were immigrants who had brought with them a rural way of living and a different social structure from the urban dwellers round them. Though they earned their livelihood from a city they remained villagers. It will be useful for us to bear in mind the northcountryman's advice about Yorkshire, 'Don't consider it as an area—it's a state of mind.'

Some urban sociologists have tried to make a definition of the city based on the criterion of size of population, especially when studying the urbanization* process in

* The word 'urbanization' does not fit easily into the strait-jacket of a definition. The demographic description is the proportion of people who are living in urban settlements or a rise in that proportion. The actual process of urbanization—the changes which may occur in social patterns when people move to towns and cities —is important to the sociologist.

regions such as parts of Asia or Latin America. In these areas some cities and towns are expanding with great speed and rapid urbanization is taking place. Breese (1969) has pointed out that it may be convenient to affix the term 'urban' to places with 20,000 inhabitants or more, so establishing an admittedly arbitrary demarcation between rural and urban by a counting of heads. However, this demarcation does suggest a number of important questions. Are those who live in settlements of less than 20,000 sociologically different from the others? Do we call them 'rural' and can we say that there will be social differences between those who live in large or small settlements? It is safer to assume that the criterion of 20,000 people may be a useful tool for studying the speed of urbanization in a particular area, but it must not be allowed to lead to a wider, unsupported theory applicable everywhere.

The fascination which the city held for some of the older urban sociologists seems to have bewitched them into uttering over-broad generalizations. Thus we find such terms as 'the urban dweller', 'the city dweller', 'urban man', 'the urban pattern', 'the urban way of life'. Moser, in his comparative study of a large number of English towns, has warned us of the need to refrain from the use of such general terms unless we are at the same time prepared to accept and to show the variations that exist between towns (Moser and Scott, 1961).

The relevance of urban sociology

For a long period, until about the seventeenth century, the population increase of the world was comparatively slow and the levels of population reached were fairly stable; then a rapid surge of population commenced which continued at an ever-increasing tempo. The process has been graphically called 'the population explosion'. Some calculations state that the present large world population will probably have doubled by the year AD 2000. Parallel with this general population expansion is an 'urban explosion'—an enormous increase in that part of the world's population which lives in cities. At present, some 370 million people live in large urban settlements of over one million inhabitants in size. It is estimated that, if this urban explosion continues, by the end of this century nearly 2,600 million people will be living in these settlements. These people would thus be a very much higher proportion of the world's population. The expansion would produce great physical and material demands and changes since vast quantities of houses, buildings, roads, communications and services of many types would be required. Those now studying social life in existing large urban areas may wonder what the future might hold in the way of intensified urban social problems.

The social changes involved would be greater still. New forms of social living could be created when men dwell in such large masses. It must be remembered, as Elisabeth Pfeil (1950) has reminded us, that man has been on the earth a much longer time than his towns and cities, which are a comparatively recent phenomenon in the history of social living; it is by no means certain that towns and cities, important as they have been in man's cultural development, are the last stage.

We are beginning to realize what urbanization implies by observing how it occasionally occurs in some relatively undeveloped parts of the world. The process of urbanization may happen with startling rapidity; the social adjustment to the new urban environment may demand great flexibility in the human beings who have to undergo this adjustment. We may wish to examine the effects of the sudden impact of urban living on a peasant who was tied to the soil and used to the social organization of a distant and isolated village, or the strains imposed on a wandering tribesman who exchanges the solitudes of the desert for the close life of crowded dwellings and the busy multitudes of the city streets (De Vries, 1961). These questions, both as they emerge today and as they will present themselves in the future, lie within the province of the urban sociologist.

Urban sociology can also be relevant in another way. According to Reissman (1965), urban studies are basic to the discipline of sociology. The process called urbanization can serve as an illustration of change in any society and the social structure and the social organization which are found in urban living can help us to understand the social structure and the social organization which are found in any society, even if not urban. Indeed Reissman affirms that some social features such as social class and bureaucracy can only really be studied in urban society. Cities and towns can be considered as a kind of social laboratory, in which a number of features of society can be studied more closely. For example, the sociological study of migration to urban areas affords useful material. People migrate to towns from less urban areas and from town to town. We may study the kind of people who move and ascertain what motives urged them to migrate—pressure from one cause or another such as the social attraction of the town. We may wish to know the social class of the migrant and whether it changes with migration or whether his residence in an urban situation leads to any significant change in social structure such as in types of family. An important aspect of our research could be the social relationship between the migrant, as a newcomer, and those who are already resident in the urban area.

One recent and valuable consequence of the growth of the discipline of urban sociology has been the recognition of the need to study urban social structure and organization more broadly and in many countries of the world. In the beginnings of urban sociology much pioneer work was carried out in Great Britain, the USA and western Europe. Some of the early theories of urban sociology were developed in those countries. This, however, had one serious defect which has become more obvious in recent years. There were special conditions in the United States and western Europe. The form of urbanization, the social history and the social structure were of a particular type often linked with a specific industrial development. It was, perhaps, almost inevitable that American, European and British sociology should show a certain outlook which has created difficulties when this sociology is applied to other parts of the world.

We know now that the conditions in these countries may not necessarily be found in other parts of the world. Speaking of Indian urbanization processes Sovani (1966) writes:

> Is the urban social situation different from the rural social situation in India and, if so, are the differences between them similar to those found in the West? ... Urbanization itself is a culture-bound phenomenon. It is unrealistic to expect the same kind of social developments in under-developed countries as in the polar types of Western cities.

It is clear that criteria which have been developed for use in the USA or Great Britain may be useless when applied to an Asian city or a Latin American town. To widen our view of social structure and social organization and to make our generalizations both more accurate and more universally applicable, we need a body of theory based on collated research made in many different countries. The discipline of *comparative urban sociology*, though still in its infancy, may supply us with new theoretical concepts which, in turn, can be applied to urban sociological theory in all countries.

Urban sociology and other urban studies

The investigation of cities, towns and adjacent or connected areas, requires a variety of different disciplines. The work of the *urban geographer* is to examine the morphology of a city; to look at its physical configuration and general appearance and to see what kinds of visual patterns are so revealed. He will want to know geographically how the urban settlement developed and in what ways it is affected, or modified by sea, lake, river, forest, hill or climate. He will delineate the land utilization and show which land is given over to houses, public services, industry, roads or leisure and what the relationship of all these factors is to the human inhabitants of the city. The *urban demographer* is primarily concerned with the population statistics of the area, its age and sex grouping, birth, marriage and death rates, and other vital statistics. He will compare and correlate these facts and study the distribution of the population. He may suffer from one limitation; in some parts of the world, accurate statistical material can be difficult to obtain. In other parts, such as some Swedish cities, he may find statistical material very readily obtainable. The *economist* will be informed about the trades and manufactures of the city and will probably have made studies that reach beyond the immediate confines of the urban area, e.g., overseas markets. The *housing expert* will know about the condition and types of domestic buildings and their distribution and costs. Other important figures, depending upon circumstances, are the *architect*, the *traffic study expert* and the *town and country planner*, whose onerous task is to study the redevelopment or replanning of an area. Occasionally, one finds the *social development officer* who is concerned with the welfare of urban people and perhaps, infrequently, the *urban historian*.

In this plethora of urban investigation we may conclude that it is difficult to outline precisely the function of urban sociology or to state exactly the relationship of urban sociology to these other disciplines. Some other disciplines may, indeed, impinge upon the borders of urban sociology. Certain of the earlier urban sociologists were willing to borrow materials from other disciplines or to employ their techniques and methods if they thought they were useful.

Zorbaugh provides us with an example of this frontier passage in his study *The Gold Coast and the Slum*. He collected demographic material (for Chicago) for suicides within a certain space of time. This data gave the addresses where the suicides had happened. He then took a large map of Chicago and marked on it, by means of black dots, the place where each suicide had occurred. When the marked map was complete, a heavy concentration of dots could be noticed in or near Clark Street. Zorbaugh had carefully surveyed this area, which lay between the wealthy Gold Coast district and 'Little Italy', an immigrant quarter, and he knew that it was a 'rooming-house' locality, where people either rented rooms or lived in lodgings; a place where unsettled people might live. Those who did not commit suicide were living in what Merton (1949) calls a state of 'retreatism'. Zorbaugh sums up the situation concisely: 'It is an area in which there can be no community tradition or common definition of situations, no public opinion, no informal social control.'

The criteria which we might apply to this passage from Zorbaugh's classic would depend very much on what we considered was the true relationship between urban sociology and the discipline of geography. We might take the view that the theory of retreatism is independently free of relationships with other disciplines. We could also take the view that Zorbaugh was right to link geography and demography with sociology but that the cart could have been put in front of the horse. Did not the urban geographer and the demographer benefit from Zorbaugh rather than he from them?

Some urban sociologists have tried to solve this problem of relationship between disciplines by proposing a different kind of approach. They suggest that urban sociology is not clearly defined enough yet to stand out sharply as a discipline in its own right; its field of research is immensely broad and complex. Therefore, the right method of study, they say, is to take a little from many other disciplines and incorporate the total material in one unified form. This 'umbrella' approach presumably implies that urban sociology is rather the queen ruling over lesser princes than the scavenger who collects what is left from the other disciplines. It has, within it, the likelihood of two dangers. On the one hand we may be so tempted by a universal, wide approach that we may lose sight of the wood for the trees. On the other hand, the urban sociologist may be overwhelmed by the material, methods, or principles lavishly coming to him from other disciplines. He may lose sight of his primary quest, the search for basic principles in urban sociological theory. R. N. Morris (1968) has given us a useful warning about the hazards of being too much influenced by data coming from another discipline. He says that it may be interesting and easy statistically to show differences between towns of 10,000, 20,000, 30,000 and 40,000 people. These statistical differences may exhibit gradings between one town and another. It is, however, improbable that we can show a similar graded change in the social groupings or in the social behaviour of people as we move from one town to another.

Urban social reform

While some people consider that cities are the noblest works of men's hands, there are other aspects to be considered. As far back as the seventeenth century, John Evelyn complained of the smoke and grime of London town. Largely as a result of social and technical changes and especially in Britain in the eighteenth, nineteenth and twentieth century, industrial and manufacturing towns developed. These were often very unlike the old towns and in them certain evils were manifest, such as diseases caused by faulty hygiene intensified by town living. In the towns large masses of people were crowded together, frequently in bad conditions. Some of these people had come to the towns, from the country, because they needed work and the town was therefore an incentive.

Mumford has used two similar phrases to describe the worst town conditions, 'palaeotechnic paradise' and 'palaeotechnic inferno' (1961). The closely built houses, in monotonous streets, were crammed with enormous numbers of human beings who had to endure overcrowding, disease, poverty and a number of other social ills. For some observers the disease, overcrowding and poverty appeared to be less awful realities than their own startling conclusions that crime, immorality, moral degradation, anti-social behaviour and deviance were appearing, under urban conditions, in more pronounced ways and more unpleasant forms. Writers were to develop the theme of urban evils from a variety of angles. Zola's portrait of the Paris slums can be compared with Upton Sinclair's condemnation of the Chicago packing yards—the 'Jungle'. Dickens wielded a powerful pen with the strength, if also the over-intensity of a great novelist. Other observers, who wrote non-fiction, varied from Thomas Chalmers, the Scottish minister turned social thinker, with wide interest in the poor districts of Glasgow and Edinburgh, to Henry Mayhew, the meticulously careful student of Londen life and the London poor (1861-1865). Another, no less vehement because of his moral polemics, was William Booth, the founder of the Salvation Army. His book *In Darkest England and the Way Out* (1890) sketches the evils of urban living and suggests remedies.

Urban poverty, however much it was sensed by thinking people, could not be accepted as a proven urban social phenomenon until the first two great scientific social surveys were made in urban conditions. Charles Booth (1902) for the great city (London) and Rowntree (1901) for the smaller town (York), carried out laborious and careful researches. They reached a disquieting conclusion; in Britain, one of the wealthiest and most highly urbanized countries in the world, a surprisingly large percentage of the population lived and died in poverty.

It was a hard task to establish the social causes of poverty, or to analyse the social evils which appear to be peculiarly connected with urban living. Social reformers at the end of the nineteenth century and later, were naturally tempted to construct a theory of social change which was linked up with improvement in human beings. The urban social reformer said poverty and other social evils were produced by the wrong kind of social environment in which crime, anti-social behaviour, deviancy, as well as mental and physical ill-health, flourished. The remedy which they suggested appeared to be straightforward and, if expensive, within the bounds of the feasible. We must alter the environment if we want to alter people and so destroy these evils. Pull down the ugly, dilapidated houses, replace them with good, well-constructed dwellings, build schools to eradicate ignorance in the children and clinics to banish ill-health and disease. Pay men a decent, living wage and their families will, henceforth, live in comfort. To young, enthusiastic people like Beatrice Webb, as we see in her diaries, this meant a golden dream of an earthly paradise.

The approach of the modern urban sociologist to urban poverty is much more complex. It must be understood that urban social reform, as such, is not a primary target of the discipline, though it may appear to be a by-product. Young and Willmott (1957) had, as their main aim in east London, the study of family structure in a city and they found that the family was affected by the move to a new housing estate. The side effect of this study was to show enlightened urban planners the need to consider family structure when planning estates. The sociologist may carry out valuable research in areas such as Deeplish,* on kinship and friendship associations which may help the town planner, but he is not merely the servant of the town planner. He may, in fact, be critical of the immense and tremendous power which can, apparently, transform a town, but at the cost of much social upheaval. The evil produced must be weighed against the good. A wider approach is necessary in observing the sociology of urban poverty. We need to be acquainted with such work as that of Oscar Lewis (1968) in Latin America and the European sociologist and town planner who are studying their latest 'slum' district should bear in their minds Mabogunje's refreshing and contrasting report (1968) on a West African city:

> It is more realistic to talk of some 70-80% of the population of the city living in virtual slum conditions ... unlike in many developed countries the slum areas are not, however, coincident with areas of moral and social deviance, criminality and delinquency. In fact life in the slum area is particularly integrated, satisfying to the full the warmth of intimate human relations. As a result, discontent with the slum conditions comes from those living outside rather than inhabitants of the slum itself.

Anti-urbanism

Among critics of urban social life were those who showed a dislike of town living generally. The later anti-urbanism had its roots in a variety of causes, but one major reason seems to have been the obvious comparison which could be made between the ugly industrial town, with its smoke, poverty, ill-health and confusion and the older, more dignified country towns where people, or at least the more well-to-do, had lived in peaceful surroundings and fulfilled some ideals of leisurely existence. Another kind

* *The Deeplish Study*, Ministry of Housing and Local Government, HMSO, 1960.

of anti-urbanism was to contrast the town with country living. Farm and village appeared to be a haven of contentment associated with the ways of living of a merrier age. In Great Britain the attitude to towns, however, showed a divergence from some continental countries. In parts of Europe the town or the city, with all its faults, was regarded as the repository of culture and the place where man ought to live in order to enjoy the benefits of civilized existence. The countryside, indeed, might be deemed the living-place of the more ignorant and the boorish where the townsman would find few kindred spirits.

The creation of the industrial town was a source of anxiety for the middle and upper classes in the Victorian period in England. Apart from the risk to health and loss of peace the towns were regarded as a threat to the established order. Their masses of working people, who had learnt both to riot and to combine together, appeared in a frightening guise; the human material of these urban surroundings seemed likely to lead to a social explosion.

There seemed to be only one way of escape—to desert the towns and go to the country. There, the threat to security posed by these dangerous, working-class crowds could be left behind at a safe and sure distance. The attitude of the country people resembled that of the middle-class townsfolk. It is reported of a certain titled country lady that she did her utmost to secure a statement from the railway company (which they could not give) that they would run excursion trains neither on Saturdays nor Sundays from a great Midland city to a pleasant health resort. 'The working class of this great city', she is declared to have said, 'do not belong to the countryside. If they are allowed to come to it, they will ruin it. They must be made to stay in the city; this is one good way to make them.'

So the exodus began. The wealthier were able to leave the vicinity of the towns completely and to move right out into the country. If they lived in comparative isolation, they were able to enjoy social contact with others in the same position. The middle classes also had the wish to move, but they were not often able to move to the same distance, since they had frequently to maintain a contact with the town in which they worked. Therefore, although some created new quarters within the town itself, for many the compromise which they achieved was the suburb. This was a district on the edge of the town, at first near to it and in close proximity to the country. The suburban dweller travelled to the town, when it was necessary, by train, by tram, by bus, by cycle and latterly by private motor car. As towns expanded and the movement outward to the suburbs continued, the distance to be travelled into the town itself became greater and greater. People clung for a long time to the illusion that they were living in rural surroundings. They still held strongly to an ideal: as Masterman (1909) puts it, 'their tiny, two-storeyed villas'—suburbia. Some people noted that this movement could, in fact, separate the rich from the poor completely, with probable detrimental results arising from this physical fissure in society, but it was not possible to draw conclusions because the working classes themselves also began an outward movement to the suburbs. If there are some urban areas where this movement is not so perceptible as in others, if may be due to the persistence of a way of housing which has lasted. The shipbuilders of the Clyde, for instance, in places near Greenock, lived so close to their work that the bows of the ships in construction appeared to tower over the tenement dwellings.

Those who were opposed to the evils of the older towns but had strong ideas about the future, were divided into two sections. One group tried to create model dwellings or, working on a broader basis, model communities in the towns or very near to them. Examples of this approach are the Peabody buildings in London or the Cadbury estate at Bournville in Birmingham. The other school of thought while wanting people to live in good dwellings, sought to accomplish their aims by building new towns or villages practically entirely from scratch. Since these communities were planned almost from the start and were relatively free from the incubi of the older towns, their creators hoped to eliminate all the faults of existing towns. Illustrations of this experimental approach can be found in Robert Owen's 'New Lanark' or Sir Titus Salt's 'Saltaire'. Owen and Salt were both wealthy manufacturers with strong interests in the working class and a staunch belief that in changing the environment of men they will be changed for the better.

Of later thinkers about the city, Ebenezer Howard has been one of the most influential. He was not so much an anti-urbanist as opposed to very large towns like London and he aimed at analysing both the faults and the merits of towns. In his *Three Magnets* drawing he stated (1946) in diagrammatic form the advantages and the disadvantages of living in town and country respectively as they appeared to him in 1898. The third part of his diagram outlines his own idea of a new town of a new type which would be without the evils of the existing towns and yet, at the same time, bring the best of country life to the town dweller. Although some of the things which Howard criticized in old towns, such as gin palaces, are not important today, the controversies which he started by the principles of his theories of town planning still have an effect in town development schemes. Perhaps the most discussed part of the theory of these early town planners is the belief that man can be changed by his environment. Anti-urbanism itself has had a great influence but in diverse directions. While some thinkers wished to change men by bringing them to 'model' towns or villages, others sought to transform the poorer areas of large cities by means of 'educational' or 'social settlements' and to create model communities. These ideas had some influence in town planning circles, but they have not created a profound respect or understanding of towns as centres of civilized living. Pahl (1968) says: 'In Britain, the most urbanized country in the world, we know so little about our contemporary towns—largely, it appears, because they have not yet been accepted.'

In this introduction we have outlined both the scope of urban sociology and its relationship to other fields of urban study and have suggested that different conditions found in urban areas have attracted the attention of urban thinkers. We have emphasized the need to consider urban sociology as a discipline of world wide and contemporary importance.

We have now to survey the different kinds of methods that have been used, or are being used, to study the social aspects of urban living and also to formulate the various theories of urban sociology.

2

Approach and method in urban sociology

The urban sociologist may view the study of his subject in two different ways. If he looks at urban sociology from the *micro-social* angle he will concentrate on one aspect of urban living, for example, the examination of an individual city, the social characteristics of urban residents in one or more urban areas, specialized urban land uses such as housing in its social situation and so on. The focus may be narrowed down to an apparently small field of vision such as the nature of a particular suburb or the relationship between people in a street and the data could be collected from field work or statistics. The study, though it may lead eventually to highly theoretical conclusions, will often have an empirical attitude. The descriptive and detailed work of the 'Chicago School' falls into this category.

The *macro-social* view, more widely based, emphasizes theory. It is concerned with the broader and relatively abstract features of urban living, with the construction of general theory and the examination of urban social institutions in a world-wide context. The questions which its exponents will ask should not be limited to any one city or one country but have a universal significance such as the question, 'What are the effects of urban living on the function of the family or on social divisions between people?' Writers such as Max Weber, who use this viewpoint, may regard urbanization as a massive and total change implying much more than the actual growth of cities.

We stress the distinction between these two views, not because they are mutually exclusive, but because it is important to understand which approach is used in any particular study. A practical illustration of research recognizing this distinction is seen in a work on middle-class families (C. Bell, 1968) which contrasts a broader macro-ecological study of towns with the smaller micro-ecological approach needed, say, in the study of a neighbourhood.

Urban sociology and urban history

Schnore suggests that urban sociology, at least in the USA, has shown, until recently, an utter lack of interest in historical questions and issues (Dyos, 1968). That this is changing is partly due to a new kind of inter-disciplinary approach in which historians and sociologists recognize the value to their own work of developments in other disciplines. This is particularly true of *social history* where certain techniques, such as sampling, are sometimes borrowed from the sociologist. Other approaches are found in the study of social structures, social groups, rural-urban differences and social contacts particularly useful for research into the family (Colloque, 1967). Sociologists also occasionally require methods extending over a period of years—'longitudinal' surveys.

The specialized study of urban history in the past has suffered from several handicaps. There was an over-emphasis on *local* history almost to the exclusion of comparative methods and the main aim of urban historians seemed to be the study of the *problems* that they found in urban history rather than the social processes which were to be observed (Handlin and Burchard, 1963) at least for cities in the nineteenth and twentieth centuries.

Cities were, in fact, considered to be *par excellence* distinctive moulds of human thought and behaviour and were regarded as the centres of special urban problems. There is now a change in historical outlook and people do accept cities, with or without their faults, as part of human existence. We do require more general studies; A. F. Weber's work (1899) is one of the great exceptions to the general rule that urban history must be micro-history. Lampard, who believes that the general study of social change is an important field for urban history, proposes that the regular routines and sequences which occur in the affairs of households and communities should be examined together with evidence of migration, occupation changes, changes of status and social mobility. The broad aim must never be forgotten. The urban historian should also consider the general study of the urbanization of society (in Handlin and Burchard, 1963).

The urban sociologist who is studying modern urban history will find the availability of materials varies. Some west European countries have good material for urban demographic studies and slightly less so for urban ecology, but it may be impracticable to obtain material for the study of social structure or social behaviour (Dyos, 1968).

The early city: the 'pre-industrial' city The evidence for the social existence of early cities is partly archaeological and sometimes uncertain in interpretation. Nevertheless, the earliest cities attract attention because they appear to involve some revolutionary change in the patterns of human social life. Certain conditions had to be fulfilled for cities to become possible realities. The old view, that development was entirely dependent upon the discovery of new technologies of food production to support city populations, has been partly modified by more recent thought, that the development of a social organization to administer the city and control the city was equally necessary.

The relationship of technology and urban social organization is still a lasting urban problem. The Aztec cities may have found that their social organization outstripped their technologies of food and water supply, transport and sanitation (Soustelle, 1955). Our modern technology of transport appears to have developed without a corresponding social organization to control its implications.

The Greco-Roman city may have had some success in dealing with these earlier problems but new and lasting difficulties were to appear. One writer says—'In every city the rich and the poor were two enemies who lived by the side of each other. One coveted riches, the other saw his riches coveted. Between them there was no relation, no service, no work which united them' (Coulanges, 1864).

An attempt has been made to classify all cities before 1750, and even some cities existing today, under the general term 'pre-industrial' which is, in fact, a 'constructed type' (Sjoberg, 1960). It is considered that these cities have some general characteristics and can be contrasted with the industrial urban cities of today. In some ways they are better known than the cities of the industrial era. Different peoples or families, as well as crafts, may be segregated, often rigidly, in conditions of congestion. Craftsmen may live, work and sell their products in the same dwelling.

The poor live on the periphery, the upper class—or elite—reside in a central position to maintain control of the social power. They may use religion as a means of control. Coulanges, for the Greco-Roman city, draws a melancholy picture of the man, exiled not merely from the city limits, but utterly cut off from the religion of the city. The elite are important to the pre-industrial city—'the more potent the elite the grander the city' (Sjoberg, 1960). They are literate and despise the merchant classes, hence the Chinese poem:

> Business men boast of their skill and cunning
> But in philosophy they are like little children.
> (Chen Tzu-ang in *Chinese Poems*, trs. A. Waley, Allen & Unwin, 1946)

Sjoberg's work has been criticized in several ways. One critic considers that he blurred the distinction between the many different kinds of city—the Ancient East, the Greco-Roman period and the European mediaeval city. He forced into the Procrustean bed of his 'constructed type' far too many cities. Cox (in Meadows and Mizruchi, 1969) thinks that the difficulty lies in Sjoberg's assumption that all cities, at least before the industrial era, were subsystems of feudal society; he points out that Brahmanic India would hardly fit into this picture. It was a system of villages rigidly organized on the basis of hereditary occupational specialization with no inherent cultural capabilities for social evolution. Athens and Rome had developed a concept of citizenship which is not compatible with feudal society. The mediaeval European city was not the product of feudalism. Lastly, feudalism and industrialism are not parallel concepts to be regarded as dichotomous to each other.

Pirenne stressed both the economic aspects of mediaeval cities and their communal organization. He insisted overmuch on the value of markets and his generalizations by no means apply to all European cities. (Handlin and Burchard, 1963) He did emphasize the peculiar character of city public life and even the co-operation that might exist—*'unus subveniet tamquam fratro suo'* (let each help the other like a brother)—but his contrast between the city life and the rural nobility cannot be strongly maintained. In some cities the *urban* nobility had much power. However, Pirenne does raise a number of important questions such as the exact nature of the relationships between cities and the country in the mediaeval period and the part played by cities in the social changes that led to the end of feudalism (Pirenne, 1956).

The city as an 'ideal type' Max Weber (1958) wished to discover a pattern in history and to find out the peculiar nature of urban social phenomena. He accepted the common idea of his day that a city was a densely populated area where the inhabitants could not know each other as well as in smaller places, but he surpassed other sociologists in his theory of the 'urban community'.

For Weber the 'urban community' was not a mere collection or aggregate of human activities, but a distinct and limited pattern of human life. In this sense the city could only appear under special conditions and at a certain time in history. These conditions were found in Europe in the pre-industrial city though Weber wisely recognized that these conditions would not be found everywhere in Europe. The actual time in history would be strictly limited.

Europe alone could produce this unique situation because of the development and influence of 'rationality'. This does not mean necessarily the triumph of 'reason' or common sense. It involves the calculated use of resources for the achievement of a particular goal or set of goals in the most economical way possible. In pre-industrial societies, on the other hand, it is claimed that there is little rationality in this sense. These societies are characterized by the converse of *rational* action—*traditional* action—which lays the emphasis on what is handed down from the past on what, it is believed, 'has always been' (D. J. H. Morgan and R. Ward, in Worsley, 1970). This difference between rationality and traditional action does seem to lead Max Weber (1947) at times into the creation of an antithesis between the two, a separation which may not, in fact, be a valid one (Parsons, 1937). Weber's lifelong preoccupation was with the development of rationality in Western civilization (Bendix, 1960). Perhaps this made it more difficult for him to think of a definition of the city which would be universally applicable.

Another difficulty is found in Weber's conceptual tool, the 'ideal type'. This stands for a theoretical stylization in which certain 'typical' elements only are stressed and it is rarely, if ever, found in practical life. If we think of the 'idea' of a dog which emphasizes all important characteristics of 'dogginess' but which cannot be applied to any particular dog, we may understand such Weberian 'ideal types' as the 'princely city', the 'plebian city', and the 'patrician city'. The ideal type may show us to what degree a particular city exemplifies this or that pattern, follows the principles of an older regime or presages a new form of organization. It may help us also to measure change.

While the ideal type was a valuable conceptual tool there are special problems which arise if we accept it unreservedly. Perhaps the most important difficulty is the way in which Weber chose the characteristics which make up his 'ideal' cities. If Weber chose his characteristics from existing cities alone, he could not really be creating a purely ideal type. On the other hand, an ideal type which was purely abstract would not have much value in understanding social situations—and understanding was a cardinal feature of much of Weber's outlook to sociology (Parsons, 1937). Friedrich says rather pessimistically 'neither the careful analyses of von Scheltung nor the thoughtful commentary by Parsons have been successful in dispelling the mist which surrounds this tool' (in Merton *et al.*, 1952). This is perhaps unfortunate as other writers were to use ideal types as well.

Weber's work, though a study in the development of urbanism rather than general urban theory, is valuable for the way in which he focuses attention on the fundamental processes of social change. Martindale (in M. Weber, 1958), states:

> Weber thought that the task of sociology was to explain human conduct in its meaningful dimensions and not merely externally. One can determine a person's loss of weight objectively by weighing him repeatedly but it makes a lot of difference if he is losing weight because he is dying of an illness or because he is engaging in ascetic practices in connection with his religion.

> The task of sociology is to explain inter-human actions in terms of the meanings they have to the parties involved as well as in terms of specific physical changes they entail.

Weber also wisely insisted on the great variety of forms of city to be found even in pre-industrial societies; while we may accept the increase and aggregation of modern populations as a possible accompaniment of the industrial revolution, we must on no account confuse the physical aggregation of population with the growth of the city in a sociological sense. For Weber, however, the end of the city appeared to be in sight. The 'urban community' had lost, or is losing, its very legal and political autonomy. The urban citizenry, no longer united by a common purpose, have ceased to man the city walls and the modern city could be in a state of decay. This melancholy conclusion to Weber's essay is echoed in Boulding's grim 'Death of the City' (Handlin and Burchard, 1963) where the older city sites are marked merely by radioactive holes.

Oriental cities Weber himself distinguished between such cities as the Greco-Roman city, the European city and the Oriental city. One writer who considers Middle Eastern and Islamic cities has underlined their contrasts with Weber's European cities. The Oriental city is in a number of aspects—legal, constitutional and material—indistinguishable from the village and its way of life. The urban locality would indeed remain a village and nothing more if it did not have the court of the ruler or his administrative headquarters (Murvar, in Meadows and Mizruchi, 1969). The Western city, however, was sharply differentiated in legal terms from other forms of social organization; the separation between city and countryside was clearly marked.

Gemeinschaft and Gesellschaft. These are two concepts which were developed in a book by Tönnies (1955). As the English translation, 'community' (or 'communion') and 'association', can be confusing, we shall use the German words. For Tönnies all social relationships are created by human wills, but these relationships can be divided into two different kinds. In *Gemeinschaft* relationships 'natural will' prevails. Peasants and villagers live in this relationship which, in its simplest form, is called a 'community of fate'. It is found in the relationship of father and child, within family and kin and between neighbours and friends. They live and work together in close contact and have understanding and concord. We have the impression that *Gemeinschaft* relationships are deep and warm and satisfying.

Gesellschaft relationships are regarded as 'separating'. Everybody is by himself and isolated and a state of tension exists as against all others. There is a concord of wills but it is created in a more impersonal way, by contract. Instead of the 'natural will' of *Gemeinschaft*, there is the 'rational will' of *Gesellschaft*.

The contrast between these ways of living is very striking. Did Tönnies intend it to be an actual picture of the differences between village and city life and did he think that an actual historical process was going on in urban areas in which *Gemeinschaft* relationships were giving way to *Gesellschaft* relationships? One modern sociologist thinks that Tönnies was historically misleading and did not know enough sociology. This critic says that the idyll of the village, at least in the past, is marred by disease, hunger, war, early death, dependence and debasement. The rural fear and scandal which beset Agnes Beaumont* in the seventeenth century are followed by the twentieth-century 'mental and spiritual destitution' which 'springs from a grave defect which was inherent in the peasant

* *The persecution of Agnes Beaumont in 1674*, ed. G. B. Harrison, Constable.

system'.* Under these conditions we can hardly say that rural life is necessarily better in social relationships than urban life, even if we accept the notion that two different sets of social relationships do exist.

Tönnies, not unmindful of the difficulties of interpreting his work, insisted that his concepts were highly artificial, even forced. They might then be regarded as ideal types. He stated (1955) that *Gemeinschaft* relationships could be found in cities as well as *Gesellschaft* types but he still believed that 'natural will' in time would be supplanted by 'rational will' and *Gemeinschaft* relationships would be replaced by *Gesellschaft* relationships.

It seems clear that the value of Tönnies for modern sociological thinking depends very greatly on how much of this thought is accepted at its face value and how much is regarded as a pointer. If we reject the historical approach and the view that village or rural social relationships do eventually develop into urban social relationships, do we also reject the view that rural social relationships and urban social relationships are different enough for there to be a dichotomy between them? The modern answer might be that if any division can be drawn between urban and rural it can only be on the basis of some practical work giving some kind of practical criteria. The second part of the answer might be that if *Gemeinschaft* relationships and *Gesellschaft* relationships exist in cities, they might in fact be found in rural areas in the same proportions. Whatever our attitude we must neither be beguiled by the goodness of a rural 'Paradise Lost' nor cloud our understanding with the view that 'the city is the home of social problems'.

It is interesting to note that so acute a writer as the mediaeval Ibn Khaldun (1957), describing cities in North Africa, stresses that it is *co-operation* that leads to urbanization. 'It is in the nature of human beings to enter into

* G. Bourne, *Change in the Village* (1912), Country Book Club, 1956.

close contact and to associate with each other.' Part of his book is devoted to the existence of group feelings in cities.

Urban ecology

The Chicago school of urban sociology Robert Park, acquainted from an early age with the problems of immigrants in the USA, became fascinated by the many aspects of social life in Chicago. During his lifetime, that city went through a period of remarkable social changes, such as very rapid population increase due to immigrants of many different ethnic and social backgrounds coming to the city in wave after wave. This growth was accompanied by noticeable social problems such as housing difficulties, crime and so on. To the sociologist this rapid change in the city appeared to be valuable; dynamic processes in human society became more evident and could be easily observed.

American anthropologists had been studying American Indian tribes. Park thought that similar methods might be used to examine aspects of city life especially 'marginal' cultures (gangs, the apparent isolation of some immigrants in the 'ghetto' and the differences which seemed to appear between people according to their residential locality). For data, both official statistical sources and private research material could be used, and great efforts were made to collect this information by the Sociology Department of the University of Chicago. M. Janowitz calls this data 'heroic ... but primitive by our standards' (introduction to Park and Burgess, *The City*). The researchers sometimes also made the mistake of confusing data and theory.

Park tried to create a theory of urban sociology. In this he was a pioneer. Knowledge about urban sociology was both fragmentary and scanty. For vocabulary and for analogies he borrowed from the science of biology. The word 'ecology' usually means, biologically, that branch of science dealing with the relationship between living things and their physical environment, together with all the living organisms within that environment. Organisms and the physical environment in which they exist are generally known as an 'ecosystem' in modern biological terms (Phillipson, 1966). The modern biologist finds it a profound and exacting task to study the complex webs of life in a tiny pond. Park's use of analogies could prove dangerous.

It seems strange that Park applied his analogies only to cities and omitted rural communities which could have been included. In fact, Park chose the city as it appeared to him to resemble the pond most closely. It had a fixed boundary, like the pond banks, since, in his view, it was socially isolated from other areas. This may have been, perhaps, somewhat truer in his time than it is today.

The analogy is developed further in the statement that the city is the 'natural habitat of man'. It acquires an organization and a distribution of population which is neither designed nor controlled (Park, 1952). Within the city, Park distinguishes what he calls 'natural areas'. These are the 'habitats' of natural groups, e.g., bohemias and hobohemias, found in large cities, 'natural' because they are not planned and are the product of forces that are constantly at work to effect an orderly distribution of populations and functions within the urban complexity. Long before Park, Engels had also insisted on the great pressures that the city could exert on the individual, 'the working-man cannot escape the present arrangement of society so long as it exists' (Engels, 1845).

Other analogies were couched in borrowed biological terms. Thus *symbiosis* is used to describe the way in which one animal lived in a common relationship with another. The rhinoceros permits the rhino bird to live in its mouth; the rhino bird, in return, picks the teeth of its protector. So men lived together in cities. *Dominance* (which in biology is a powerful tree or plant dominating an area) was applied to a firm or firms dominating a town. *Invasion* by insects, plants or animals could drive away other species which would then be succeeded by the invaders. The analogy was too tempting to resist. Was not this the process that some Chicago areas went through as immigrants 'invaded' them, driving out the original residents?

Modern attempts at interpretation of the way in which people live in different districts in towns usually begin from a different angle. Thus one geographer writing about West Indian migration to Britain says 'there is a universal and probably correct assumption that ghettos are the geographical expression of social failure' (Peach, 1968). Jones (1960) shows the segregation of Roman Catholics and Protestants in a Northern Ireland city varied with the degree of tension. Firey found that a group of Italians who lived in dilapidated and decaying property, refused to leave the area and move into better houses, though financially capable of doing so, because they were strongly attached to their houses through cultural and ethnic values (Reissman, 1964).

Park's solution of the difficulties created by biological analogies was to divide human social organization (unlike animal biology) into two levels—a higher, or 'cultural' level and a lower or 'biotic' level. His problem was to explain the relationship between the two. Either the biotic level had to be independent of the cultural level or it had to reflect accurately what went on at the cultural level (Reissman, 1964). Neither was true. It seems that social life more closely resembles the croquet game in Alice in Wonderland than it resembles a biological organism.

Further developments in urban ecology: critical attitudes

The 'natural' area Although Park considered this to be an important part of urban study, he never really offered us an exact and standard definition. The natural area is defined as a geographic area the characteristics of which are determined by unplanned, impersonal, subcultural forces. The slum, the ghetto, the industrial complex and the central business district, were unplanned (Reissman, 1964). 'It is assumed ... partly as a result of selection and segregation and partly in view of the contagious character of cultural patterns, that people living in natural areas of the same general type and subject to the same social conditions will display, on the whole, the same general characteristics' (Park, 1952). One aim of urban ecology, therefore, was to identify the natural areas of the city, but Park tended to use certain terms indifferently. He used 'natural area' and 'neighbourhood' as if they were the same thing and also confused 'natural area' with 'community' (König, 1968). Park's followers themselves do not agree about the precise meaning of 'natural area'. Zorbaugh considered the natural area to be essentially a product of land utilization while McKenzie, on the other hand, defines it in terms of certain characteristics—population, race, language, income and occupation. Burgess came nearer to a synthesis of these two opposing views. He said that community life is conditioned by the spatial distribution of the population. In this sense community must be understood to refer to three distinct aspects of life—ecological, cultural and political. We must not forget that the Chicago ecologists tended to assume that every community could be broken down for analytical purposes into distinct areas; the difficulties here would be that the boundaries might be arbitrarily drawn or the areas chosen themselves might be of no great value.

Let us suppose that we wish to study the incidence of mental disorders in a particular city; it might happen that our researches would show that the occupations or social class of the people studied are more significant factors than their residence. Alihan has added another relevant and trenchant criticism of the ecologists. She thinks that they concentrated far too much on the 'externals'—railways, highways, factories—the physical environment—to the neglect of other important factors.

Another general criticism of the Chicago school is that they looked far too much at the big cities with their 'special problems' and neglected regional centres or small towns, nor did they seem to consider the need for comparative study of towns (Robson, 1969; Burgess, in Park and Burgess, 1925).

The zone hypothesis The Chicago ecologists left us other concepts which have proved suggestive in urban studies. The attempt to analyse urban spatial structure is reflected in Burgess's hypothesis—the concentric zone theory (Burgess, in Park and Burgess, 1925). In diagrams this shows a small inner zone—the central business district surrounded by the decaying buildings of Zone II—the zone in transition. Beyond this lies the zone of the working people—Zone III—which is surrounded again by the Zone IV—the better residences. Outside the circles of the city lie the commuter area or dormitory suburbs. Although Burgess drew the diagram originally for Chicago and perhaps thought of applying it to other American cities, it is only right that we should remember that it is a way of considering the facts and not a representation of the facts themselves. Indeed, many cities, including Chicago, do not conform to this 'ideal' pattern exactly and many Latin American, French or South-East Asian cities might not conform at all. It is not an accurate general statement to say that cities expand from the centre outwards. However, there are two important legacies which are left to us from this theory. The first is the zone of transition, which deserves special treatment. The broader legacy is the necessity to study the relationship between urban spatial structure and the social organization linked up with it (Hauser and Schnore, 1965).

The work of the Chicago ecologists did not actually produce a total theory of the city but the importance of ecological studies is still with us. Madge suggests that field studies should be used to provide reviews of ecological changes in a city (Madge, 1963). Specialized modern ecological approaches such as T. Morris (1958), Schulz (1969) and Schorr (1964), show a concentration less on areas *per se* than on people and their social relationships. They should be compared with older studies such as Wirth—*The Ghetto* or Zorbaugh (1929). Schnore also pleads for ecological studies on a cross-cultural and historically comparative basis in Meadows and Mizruchi (1969). The urban field is a battlefield between those who stress the impact on urban life of 'objective conditions'—the external environment, population structure and the like—and those who emphasize the role of social or cultural values as a key determinant of the so-called objective conditions of human life in general (Sjoberg, in Hauser and Schnore, 1965).

Cross-cultural studies can be valuable. Quinn has pointed out that certain assumptions underlie the Burgess hypothesis. These are the existence of a particular kind of economic basis and economic and cultural factors such as the private ownership of property, together with a value orientation which assumes the motive of profit maximization. We may well ask, 'But what happens when these assumptions are missing?' Pioro's (1962) study of two Polish towns has maps showing demographic details on a spatial basis (age, sex, occupations, immigration, density of population, diseases, sewage facilities, etc.). Other maps show the spatial distribution of crime, housing and even drunkenness. Pioro comments, 'The territorial structure of a town does not depend only on its geographical relief ... it is to a great extent the result of spontaneous social forces whose territorial relationships are recognized by social ecology.' Pioro compares the development of a town under capitalism with that under a communist regime. He asks, 'What should be our hierarchy of needs in town planning? Do we put children higher than motor cars, schools than level crossings?'

The Shevky technique: social area analysis

Thirty years after Park's *The City*, Shevky introduced his method of studying social differentiation and 'social space' (Shevky and Bell, 1955). For Shevky, 'urban social structure' can be examined under three basic 'constructs'.* 1. *Social rank* (occupation, education and rental). 2. *Urbanization* (fertility, women in the labour forces and single family dwellings). 3. *Segregation.*

A high ranking under social rank will be shown, for example, by a high proportion of non-manual workers, of persons with long formal education and high rental levels. Segregation is a measure of the concentration of ethnic or minority groups in a given area in relation to their proportion in the urban area as a whole. The scores of these variables can be allotted positions in a two-dimensional 'social space' diagram. The technique could probably be applied to country as well as to a town. Shevky neglects ecological space and focuses on social differentiation.

The basic criticisms of this technique centre upon the constructs.

Social rank is not unfamiliar, corresponding to 'socio-economic' status. 'Urbanization' is close to the ideas of Wirth and suffers from the same criticisms. However, as Robson says, 'there is no logical procedure whereby the selection of the particular indices can be justified ... there appear to be a variety of possible derivations in place of those selected by Shevky' (Robson, 1969; Timms, 1971).

Multivariate analysis

This is a method of *comparison* within a country or per-

* 'Constructs' are theoretical terms, they are not observable ... In order to relate to empirical material the meaning of a construct has to be translated ... into operational procedures' (Timms, 1971).

haps even between countries. A British example is Moser and Scott (1961) which applies statistical methods to urban sociology. The resulting typology presents specific 'components'—social class, population change, overcrowding and so on. J. Wakeford (1968) has also suggested simple and practical methods for the construction of scalograms of urban phenomena for urban comparisons.

General typology of urban developments

Reissman (1964) has made typologies showing societies compared in certain graduated scales. The forty-four countries chosen are grouped finally into a total typology of urbanization and he discusses the different stages through which countries may appear to pass.

Comparison of rural and urban areas by statistical analysis

This is yet another 'classification' method, but it presents a number of considerable difficulties largely stemming from the problem of obtaining exact division between the areas 'rural' and 'urban'. In the USA, Duncan and Reiss (1956), have assembled a large amount of data based on the US Census returns, but their conclusions are that: 'there is no sharp dividing line between urban and rural communities, either in spatial terms or in terms of community size and functional specialization.' In the USA it is thought that the largest urbanized areas display a closer resemblance to the 'rural-farm' category than to the smaller urban places.

The methods which we have mentioned so far depend chiefly upon quantitative techniques which in themselves raise special difficulties. Morris says: 'All schemes for classifying the cities in urbanized societies have had to resolve the problem of devising categories which will be clear and unambiguous without becoming over-simplified' (R. N. Morris, 1968). The problem is twofold; we have to choose the right kind of statistics and to discover factors which can be displayed or demonstrated by statistics if at all. Morris has examined critically classificatory schemes.

It is most important for the urban sociologist not to be beguiled overmuch by the Circe of statistical presentation. In some parts of the world it is difficult and sometimes almost impossible to obtain rigidly defined statistical material; other methods of obtaining information must be used. Also it is especially essential that we look at the qualitative aspects of urban living. How could we analyse statistically de Lacretelle's saying,* 'The city has a face; the country has a soul'?

The Redfield-Wirth approach

Redfield Redfield was an anthropologist who carried out work chiefly among villages and small communities in Mexico. His field work had the aim 'to define differences in the nature of isolated homogeneous society on the one hand and mobile heterogeneous society so far as these kinds of society are represented' (Redfield, 1941). He obviously implies that there are two kinds of societies and that the difference between them might be defined in some way. The first society could be called 'folk society', the second 'urban society'. His main variables showing change from folk society to urban society were: the increase in cultural disorganization, the increase in secularization and the increase in individualization.

Redfield was trying to create a 'continuum' or scale. At one end would be an 'ideal' type with all the essential characteristics of a 'folk society'. At the other end of the continuum—the polar opposite—would be 'urban society'. Unfortunately Redfield did not define 'urban society'. In different places on this continuum would be the communities which Redfield was studying. It will be noticed

* Quoted in *Readers' Digest*, December 1969, p. 89.

that there is implied a continual gradation from one pole to the other of the continuum.

A penetrating, though friendly, criticism of Redfield highlights the difficulties of his theory. O. Lewis (1951) suggested that the folk–urban concept of social change focuses attention primarily on the city as the source of change to the exclusion or neglect of other factors. He thinks that *conflict* actually was a significant part of Tepoztlan life; there were definitely tensions in this village. Redfield seemed unable to accept the 'disorganization' of city life yet some 'disorganized' cities have shown a remarkable degree of permanence. A further criticism is that the folk–urban constructs should be used as research tools, not as generalizations, that they do not necessarily represent different stages in the evolution of societies and that they are the invention of Western writers (Hauser and Schnore, 1965). Pahl argues for the reconsideration of the rural urban continuum on a temporal rather than a spatial basis (Pahl, 1968). His article draws attention to the need for comparative work and wider perspectives. Barclay, for example, has a delightful picture of the Sudan *villager* returning from work in *urban* Khartoum and dropping his urban ways as he changes into his cool, village *galabiah* (Barclay, 1964) and Janet Abu Lughod shows us Egyptian villagers pouring into urban Cairo and 'ruralizing' the city (in Breese, 1969).

Louis Wirth Wirth—perhaps the most controversial figure in urban sociology—was involved in the Chicago school. His *The Ghetto* has become an early ecological classic. He suffered from the underlying pessimism about city life which afflicted some Chicago sociologists. He also lacked wider information about cities in other parts of the world. Urban sociological studies of Africa, Latin America and South-East Asia were then in their infancy.

His approach to urban studies is partly 'historical'. For him the beginnings of civilization were signalled by the growth of great cities. The shift from a rural to a predominantly urban society which has taken place in such industrialized areas as the USA and Japan has been accompanied by profound changes in virtually every phase of human life. The city, however, is the product of growth rather than instantaneous creation; the influences which it exerted upon the modes of life should not be able to wipe out completely the previously dominant modes of human association. 'To a greater or lesser degree, therefore, our social life bears the imprint of an earlier folk society ... the community and the country may be regarded as two poles of reference to one or the other of which all human settlements tend to arrange themselves' (Wirth, 1964).

The definition of the city Wirth selected for his definition, those elements of urbanism which marked it, for him, as a distinctive mode of life. The city is 'a relatively large, dense and permanent settlement of socially heterogeneous individuals.' The sociologist had to discover the forms of social action and organization that typically emerged in such settlements. Wirth's definition obviously contains variables; the larger, more densely populated, more heterogeneous a community, the more accentuated the characteristics associated with urbanism will be.

The effect of the city on human beings As the city grows in size, with more residents, the bonds between residents will grow weaker and the character of social relationships will alter. They will become 'impersonal, superficial, transitory', and the city dweller will treat relationships in a 'rational, sophisticated manner' (i.e., for his own ends). Contacts with people will be close but superficial and life will be lived at a fast pace. This resembles St Paul's Athenians. 'They spent their time in nothing else, but either to tell, or to hear some new things.' The size of the city means that indirect communication is necessary; all the people cannot meet together and there will be economic changes such as highly developed division of labour, the domination by large firms of small businesses and the competition for a scarce resource—land. Wirth stressed the variety of groups in cities, the conflict of loyalties and the greater geographical and social mobility of the city dweller.

Criticism of Wirth Wirth himself realized that there were certain definite limitations to his work. He did not suggest that urbanism as a way of life was peculiar to city dwellers and he accepted that small-town residents might show more urban patterns. He also recognized that large cities have been built with little capitalistic or industrial development; and he did not try to classify cities.

The 'key variables' The critic must prove that some important characteristics of urbanization cannot be explained in terms of the three minimum key features. Studies of India show that some Indian villages in sheer density and heterogeneity can almost exceed the city of Calcutta, but this has not brought about a cultural transformation (Turner, 1962). Further, there is no reason to suppose that cities everywhere have played or will play the dynamic role that Wirth associates with them in the West.

Wirth obviously tries to compare urban society with some other different form of social life—'the folk society'. He assumes that rural social relationships tend to be primary in character while those in urban society tend to be secondary; the very variety of social relationships found in cities contradicts this view. There are primary relationships and even 'villages' in cities (Young and Willmott, 1957, Gans, 1962, Abu Lughood, in Breese, 1969). It seems also possible that Wirth's error may have partly occurred because he compared only one part of the city (the inner area—especially the zone in transition) with rural areas and that he did not consider other parts of the city sufficiently nor did he study the stages through which city residents could pass.

Some of Wirth's attitude to the city may have come from Simmel (1956; Spykman, 1966). Simmel says 'The deepest problems of modern life derive from the claim of the individual to preserve the autonomy and individuality of his existence in the face of overwhelming forces.' Simmel describes the reserve of 'metropolitan life which in the eyes of the small-town people appears to be cold and heartless but is, in fact, a kind of protective mechanism.' Wirth paints a picture of the 'rootlessness of the city—personal disorganization, mental breakdown, delinquency, crime, corruption, disorder'—as more prevalent in the urban than in the rural community. Janowitz, however, in a practical study (1952), considers that a significant proportion of the metropolitan residents are not 'rootless' individuals and that counter-trends to large-scale organization continually develop which modify the effects of technological impersonalization. He thinks that earlier sociologists, and this could include Wirth, were 'marginal' people, who exaggerated the 'rootlessness' of urban existence and he draws attention to the importance of the local community.

Schelsky (1965), in an article called 'Is the city resident really lonely?' suggests that the city dweller prefers the *city* rather than the small town or village. This does not apply to the recently arrived migrant who may feel lonely during a period of transition. The city dweller, in his free time, wants no 'neighbourhood' relationships imposed on him and desires to pick and choose his neighbours. Schelsky, like Wirth, is really thinking mainly of Western cities.

3

Urban social structure

The term 'social structure' has not yet acquired a precise and rigid degree of definition in some sociological writings. It may be found in a variety of usages; to denote a social regularity, of perhaps a repetitive order, not merely a random arrangement. It may also be used as a general expression to cover the broad arrangement (units, sub-systems, institutions) as in the title *The Changing Social Structure of England and Wales* (Marsh, 1965). One group of sociologists assume under social structure the social integration of a society and its sub-systems (structural–functionalist approach). From this point of view, deviation could occur between culture structure (goals) and social structure (means).

However, it might seem wise when approaching urban social structure, to use initially this statement, 'human beings are connected by a complete network of social relations. I use the term social structure to denote this network of actually existing social relations' (C. R. Bell 1968). Beshers asks, 'Is there a distinctly urban social structure ... a set of persistent patterned social relationships among persons or positions?' (1963). It seems clear that we have to show that there is something special about urban social structure that makes it recognizably and significantly different from the social structure of our society as a whole (Mann, 1965). We might even have to show that if changes in urban social structure occur, they are changes that are peculiar to that structure alone and not connected with any wider changes in society.

It will be noticed that there are several variations in the method used to study urban social structure. Some of these combine an ecological approach with other aspects in a greater or lesser degree. Some modern urban ecologists who believe that the only unique integrant of urban sociology is urban ecology, insist on the importance of human spatial distribution in and around cities as the only major urban phenomenon that remains sufficiently separable to provide a legitimate intellectual rationale (Thomlinson, 1969).

Beshers has used, among other material, the study of marriage patterns combined with the effects of 'social distance' to produce his own theory of urban social structure. He would say that within the cities in the USA which he studied there are several systems of social structure and he distinguishes between the 'urban stratification system' which is shared by the inhabitants of the city and the numerous sub-cultures within which we investigate actual concrete social relationships (1963). Extensions of this study would show, probably on a time scale, actual changes occurring, such as changes in the social pattern of housing, and would accept that there is a reciprocal relationship between the spatial structure and the social structure, between physical distance and social distance. Segregation of the poor in a slum, for example, will have different sociological consequences from their being scattered more randomly throughout the urban area (Pahl, 1970).

A second facet of the study of urban structure is that urban life may exhibit sufficient regularities for us to extract systematic inter-connections which we can arrange to exhibit a structure and we can study how this structure changes. Since we are examining a structure of social relations, we know that it has to be analysed in terms of roles. This is the starting point of Gluckman's analysis of tribalism in African towns (Southall, 1961). If we use this approach we may concentrate especially on the problems of sustained growth which are created in towns or through towns by modernization. The structural characteristics of modernization have been identified as the development of a very high degree of differentiation. The Havasupay Indian who prayed 'Sun, my relation ... be good and shine ... do good deeds for us humans. Let me remain always as I am now' could be contrasted with the modernized, dynamically oriented life of a person who wants to be always different—the city dweller (Eisenstadt, 1970). Attempts have been made to measure the development of this differentiation (G. and M. Wilson, 1945) and have come under some critical fire (Timms, 1971), notwithstanding which, the study of African towns will be important as showing changes from small scale to large scale social systems and as illustrating the need for comparative studies of urban social structure.

Urban living and the family

'The human family constitutes the beginning and the essential element in society' (St Augustine, *The City of God*). The family is a very important human group in urban social structure and a precise definition is necessary.

(1) *The nuclear family*. This unit consisting of husband (father), wife (mother) and their (immature) children, is found universally. Its importance, however, varies in different societies, depending on the position of the nuclear family itself in the wider field of kinship and the function it fulfils. Sometimes kinship affiliations appear to overshadow the nuclear family, sometimes the nuclear family bulks more largely than these kinship affiliations. A married individual, with children, belongs to *two* nuclear families, his own family of *orientation* (in which he was born and reared) and his family of *procreation* (his wife and children). These two families give rise to kinship systems

and societies are sometimes classified according to the types of kinship they emphasize. It could be that the study of kinship in urban industrial societies may have been somewhat neglected in favour of kinship studies in simpler societies (Bottomore, 1962). Important functions of the family can be the sexual function, the reproduction and socialization of young children and sometimes, in the past or present, the economic function.

(2) *The extended family*. A wide definition which clears a track through the 'jungle of terminological confusion' (Rosser and Harris, 1965) is that of Bell and Vogel (1960) 'any grouping, related by descent, marriage or adoption, that is broader than the nuclear family'. This definition considers the extended family as a separate social entity and separates the concepts (sometimes confused) of household and family. There are a number of forms of the extended family. This fact is especially important because of the assumptions frequently made about the 'effect' of urbanization upon *the* extended family. Baric has pointed out that if by extended family is meant a group of kin of *three generations* or more with a fairly well-defined corporate life involving co-operation in productive activities, such units are almost entirely absent from modern urban society (Firth *et al.*, 1969). If we mean the extended family in a much looser sense which can mean the members of a kin who maintain a relationship of some intimacy with members of a nuclear family, then such entities do exist in fully developed urban conditions.

Much of the literature concerning urbanization and the extended family deals with change and there are two assumptions—Wirth's is, perhaps, the most pessimistic assumption. 'The family appears to have lost so much in urban society, that the reader may wonder whatever it has got at all' (Mann, 1965).

The other assumption is that the nuclear family is replacing the extended family in importance in urban industrial societies because it is a better 'fit' in urbanization.

In its extreme form, this is the view that the extended family is in fundamental conflict with a modern industrial economy. Historically, it is not certain that there has been a change from the one type of family to the other, at least as a result of urbanization. Indeed, it is possible that close family relationships might have developed as a result of industrialization and urbanization because the family represented both security and defence. The rickshaw-wallahs of Lucknow, a poor, homeless and dispersed group, showed strong kinship solidarity and control in the face of hard urban conditions (Piddington, 1965). The real tragedy in urban India comes for those without kin in the city, such as a destitute and elderly village couple.*

Some, however, would say that disintegration of family types does occur, but is only a temporary phenomenon. Others completely deny that urbanization causes any disruption at all and would point to the vigour and vitality of the extended family shown in a number of areas. Lastly, there are those who argue that a modified form of extended family has emerged as an adaptation to changed circumstances. Goode (1963) believes that there is a general world wide change occurring towards a conjugal system, but that the changes occur too widely to be attributed solely to either industrialization or urbanization.

It is true that industrialization requires a *mobile* labour force which might interfere with kinship ties but this should not be overstressed. Part of the folklore of sociology has been that the extended family hinders geographical mobility (Cotgrove, 1967). A more important facet of urbanization and the family is the way in which individual members of the family may be affected. In Asia and the Far East, the role of the woman may change (Hauser, 1957; Ross, 1961), sometimes drastically, so that she may become the pivot round which everything revolves. This could lead to a diminution in the authority of the husband. The pivotal position of the mother has been studied in Wales

* See Kamala Marka's *Nectar in a Sieve*, Putnam, 1954.

(Rosser and Harris, 1965) and in England (Willmott and Young, 1961, Young and Willmott, 1957).

The importance of kinship Even in non-urban societies there seems to be some ambivalence about the relative importance assigned to kin by individuals or families. C. M. Doughty was called in, because he was a *complete stranger*, to mediate in a dispute between two Arab kinsfolk.* In Oxford, blood ties formed a relatively minor part of the total activities of an old established urban society since acceptance or rejection of kinship ties is permissive in England. Neighbours, however, were in a different category—they had to be handled with great care (Mogey, 1956). In South and Central America a peculiar form of artificial kinship has been noted which carries obligation—the godparent relationship; 'compadres' (Whitten, 1965; Lewis, 1968).

The family and urbanization can only be studied if we widen the meaning of 'family' and look on it as a process which is occurring in the lives of individuals. It is valuable to examine the relationship between the different generations (Townsend *et al.*, 1968).

Social interactions in urban societies

The network concept From the study of kinship, there has been a development into other fields of urban study. Bott (1957) analyses in urban areas, the variations that occur in conjugal role performance. Role-relationships for Bott are those aspects of a relationship that consist of reciprocal role expectations of each person concerning the other. The variations are related to the form of the family's informal social network; they are associated with the pattern of relationships among these acquaintances and kin themselves. Her hypothesis is that the degree of 'segregation' in the role-relationship of husband and wife varies directly

* C. M. Doughty, *Passages from 'Arabia Deserta'*, Cape, 1931.

with the connectedness of the social network of the family. It is possible to arrange families on a continuum ranging from those having highly segregated role relationships to those having joint role relationships. There is an interesting contrast between small-scale, more isolated, relatively 'closed' local groups frequently encountered in primitive societies and in certain rural areas of industrial societies. Here segregation of conjugal roles is likely to be even more marked than for urban families with close-knit networks. Moreover, she finds that urban families are more highly *individuated* than families in relatively closed communities, i.e., the elementary family is more separated; the difference between an urban family and a family in a relatively closed community is one of degree.

Criticisms of Bott are more of method than of aims. One critic thinks that Bott has relevance only for middle-class couples of a certain age group (Udry and Hall, 1965), another that more refined sampling techniques are needed (Aldous and Strauss, 1966). Harris (1969), while stating that Bott's study needs greater conceptual clarity, says, however, 'she was on to something important, but failed to state its precise nature.'

The general study of networks What Bott was 'on to' was the study of networks. Mitchell has wisely pointed out that Bott's study should not be considered merely as associating social networks with conjugal roles (Mitchell, ed. 1969). There appear to be three different orders of social relationships which are characteristic of large-scale societies—possibly of all societies—but particularly of urban social systems. These are the *structural order* by means of which the behaviour of people is interpreted in terms of action appropriate to the position they occupy in an ordered set of positions, such as in a factory, a family, a voluntary association, a trade union, a political party or similar organization. The *categorical order* by means of which the behaviour of people in unstructured situations may be interpreted in terms of social stereotypes such as class, race, ethnogeny. The *personal order* by means of which the behaviour of people in either structured or unstructured situations may be interpreted in terms of the personal links individuals have with a set of people and the links these people in turn have among themselves with others. These are not three different types of actual behaviour; they are rather three different ways of making abstractions from the same actual behaviour.

One difficulty in the network concept is that sociologists have not yet reached agreement on a common set of features for all networks. Some features which are important are density of relationships (Bott, 1957), the number of persons in contact with the 'anchorage' person (the person being studied) and the durability of networks. Some networks only come into being to deal with crises and to create communication channels. Frankenberg (1966) uses the words 'social redundancy' to describe the spare channels required in social interaction of people, so that if one channel is cut off another can take its place. One idea is that people who are bound together in many ways are more securely bound. He thinks that modern communication theory can be profitably applied to the study of social changes leading from rural community to urban alienation and anomie. Communication can also be studied through networks to appreciate norms. Bott (1957) says

> When many of the people a person knows interact with one another, that is, when the person's network is close-knit, the members of his network tend to reach consensus of norms . . . when most of the people a person knows do not interact with one another, that is, when his network is loose-knit, more variation in norms is likely to develop in the network and social control and mutual assistance will be more fragmented and less consistent.

Mitchell (1969) advises that the analysis of social relationships in terms of social networks does not mean a substitute for an analysis in terms of social institutions. An institutional analysis utilizes a partial network to erect a logically coherent structure of norms and behaviour patterns (e.g., the kinship system). A single aspect of the complexity of human behaviour is considered at a time. The network approach, however, deliberately seeks to examine the way in which people may relate to each other in terms of several different normative frameworks at the same time.

Social difference and urban living

'We are now entering a "popular" quarter. Only poor people live here and, as you can see, it is not pleasant. I prefer my own home in another part of the big city. We pay a high rent but living there is better for us. We do not meet these people very much' (an Italian urban lady).

This highly subjective and personal remark revealed certain social differentiations in a big city. Poor and rich lived in a measure of segregation from each other. The poor, at least in the opinion of the speaker, were worse off spatially as well as from an economic point of view, since they could not afford the high rents required to live in a better part of the city. The element of segregation is accepted as, to some extent, is the view that some people are better than others. I emphasize the subjective nature of these words because subjectivity does occur in situations where one might have wished a more objective view. Jones (1960) speaking of Belfast says

> The concept of class is central because it reflects so many attitudes of the people involved. A person will often identify himself with a particular class ... and will act according to his concept of expected behaviour of the particular class ... such matters as the part of the city in which he wishes to live reflect to a large extent his own estimate of his particular place in society.

In Australian towns it was observed that a higher percentage of the 'workers' considered themselves to be 'middle class' than in England (Oeser and Hammond, 1954). We must remember, however, that with certain limitations, class reflects the ability to choose where one lives. Carolina Maria de Jesus said, 'I can't give my children a decent house to live in ... I'm trapped in the life of a *favelado*' (*Beyond All Pity*, Four Square Books, 1962).

Social stratification This term is sometimes used to describe the placement of people into a social category for sociological study. The 'layers' are not the same everywhere, or at any period of time. The persistent and rigid stratification of caste in India, which restricted movement closely, has not yet been studied in urban conditions where change might be expected (Bottomore, 1962). The mediaeval 'estates' society of Europe also restricted movement, though not so rigidly, and strata were distinguished legally and by work—'Those who fought, those who prayed, those who worked (i.e. with their hands) (Dupont-Ferrier, 1929). The difference between these societies and modern urban industrial societies is considerable and calls for some historical sense in interpretation. Some sociologists accept a historical development of 'classes' and 'class conflict' in the early period of industrial capitalism but believe that this has ceased in the fully evolved industrial societies. This is an assumption which needs some correction from recent studies of the evolution of social classes, social movements of the present time and the possibility of future social change (Bottomore, 1962). Foster studied some Victorian towns to ascertain what 'class' meant in the nineteenth century (Dyos, 1968) and to see how far it was actually perceived by the people of that time. Thompson (1968) says, 'class happens when some men, as a result of common experiences (inherited or shared), feel and articulate the identity of their interests as between themselves and as against other men whose interests are different from (and usually opposed to) theirs.'

The two historical thinkers, Marx and Weber, offer

different approaches. Marx considers really only two main 'classes', the 'bourgeois' and the 'proletariat', each conscious of its class position and in conflict with the other: 'The history of all existing societies is the history of class struggles.'* For Marx, the towns were very important. He speaks of 'rural idiocy'. Weber uses 'social status' but in two senses, 'class status' (economic position) and 'social status' (in the sense of prestige, usually visible in a particular style of life). Possibly neither Marx nor Weber can offer a completely satisfactory practical explanation and researchers often use a statistical rating such as occupations placed in a scale, the socio-economic levels found in the Government Census or the Hall-Jones scale. The distinction between class as used in an economic sense and prestige in the sense of life style should be maintained.

The evaluation of social differences in urban areas We might find it valuable to begin with social differences in rural areas. They could be considered as simpler, or less flexible than differences in urban areas; the urban dweller might consider them as far too rigid. In one East Anglian village the hierarchy of status or prestige was clearly marked; it was understood that the post-office lady had married beneath her and gone down one of the *nine* grades by her action.

The same differences appear to apply, though not so closely, to small-town class structure, which is affected by certain factors such as the limited size of the community, giving residents more intimate knowledge to form judgments of relative status. Strata are less numerous than in urban areas and lines of division are more final and permanent with a higher degree of consensus about the standards of differential evaluation and social position (Boskoff, 1970). We have to show caution about early research into urban status systems (e.g., that of Warner,

* K. Marx and F. Engels, *Manifesto of the Communist Party* (1848), M. Lawrence, 1934.

1949) in small, slowly changing areas. Conclusions were produced which could not be applied to larger, more heterogeneous and dynamic areas.

Class consciousness and urbanization: the evaluation of social differences It is, however, difficult to evaluate the status of all residents in an urban area, partly because the factor of segregation tends to blur otherwise meaningful connections between the city or region as a unit and the activities of numerous specific persons and families. Touraine, who has studied working-class consciousness in towns of varying size in France (1966), found that there was a tendency for it to diminish with the size of the town. In Paris, this was partly because of the better and more diversified life chances offered by a great city. In Zola's *L'Assommoir*, the working-class guests were able to visit the Louvre after their wedding breakfast!

However, Touraine points out that the concentration of masses of working people, with a low standard of life, in certain districts has led to the formation of class-consciousness—so far as this applies in urban areas.

How far is the development of class-consciousness actually linked up with the process of urbanization? Western sociologists, who have been accustomed to regard urbanization as something linked up with industrialization, tend sometimes to regard the development of industrialization as an essential means of improving living standards of 'under-developed' countries. It is not certain that this would necessarily happen nor does the impact of industrialization of itself lead to the development of urbanization. We should remember that most 'industrial' cities in western Europe as well as in the USA are comparatively recent and they had some characteristic features, a tradition of self-government and a demand for labour which was associated with their growth. People moved to cities because opportunities for work were found in them. These cities also played a role in the process of cultural change rather than of cultural innovation and despite episodes of severe social disorganization they were able to overcome disruptive tendencies through these characteristics (Hoselitz, in Hatt and Reiss, 1957).

Urbanization and less developed countries

In comparison with the advanced countries, there are hardly any industrial cities in underdeveloped countries. The older cities which exist in these countries do not support the generalization that a city is the readiest place for innovation. The process of urbanization in these countries merits special attention. In Africa, while there is no one fixed pattern of African urban life, there is a 'type' of urban African; one of the features of some Copperbelt towns is the way in which 'tribalism' continues to exert a powerful force on African urban life even in situations where it would seem to have no place. There is an emerging class structure in the towns which is beginning to show lines of division of urban communities into economic classes. Even here tribalism may appear to set the lines within which these divisions come. Mitchell's example of the tribal Kalela dance is considered to show the increasing significance of prestige or 'class' as a category of social interaction in African urban life (Epstein, 1958).

In Latin America, where there has been a very swift growth of cities without a country-wide growth of rapid industrialization, there is a rift between the urban 'classes' and the rural 'masses'. The advanced cities, therefore, ignoring the rural masses, are the sources of power but rest on a precarious social structure with the upper social stratum holding on to the reins of power. The landed aristocracy, no longer so feudal, provide primary goods for the national and international markets and are not separated from the upper classes of the cities (Horwitz, in Beyer, 1967). The rift becomes deepest when the landowner clashes with the disenfranchised and inarticulate rural masses in such a situation as that described in Jorge Icaza's stark novel* where the landowners and the gringos combine to destroy even the poor huts of the Indians. Here control of the city, probably the leading city in the country, means control of the nation.

Urban social differences and segregation If we accept that social differences do exist in towns, we may also find that in larger towns and cities there may emerge patterns of physical separation linked in some way with these social differences. At Oxford, differences in education were the most significant indicator of residential segregation (Pahl, 1970). At Dagenham, the residents were mainly working-class manual workers (Willmott, 1963). The kind and degree of segregation may vary greatly but it will suggest a number of important considerations.

One of these is the idea that people are separated, because of social differences, either voluntarily, compulsorily or accidentally and they want to remain separated at least in their housing. The celebrated Cutteslowe Walls in Oxford were apparently built partly because the people in a private housing estate in the early days disliked having council-housed people as their neighbours and this dislike was rooted in a consciousness of social differences between the estates (Collison, 1963). While the children mixed well, their elders may have thought differently.

> My parents kept me from children who were rough
> Who threw words like stones and who wore torn clothes.
> Their thighs showed through rags. They ran in the street
> And climbed cliffs and stripped by the country streams.
>
> Stephen Spender, *Collected Poems*, Faber, 1955

Or was the dislike due to a not unknown fear of loss of social position? 'To live on a Council estate is social death' (quoted in Mogey, 1956).

* *Huasipungo*, trans. M. Saville, Dobson, 1962.

This Oxford incident poses an issue which recurs constantly in town planning discussions. If we assume that town planners have power to determine where people shall or shall not live, should a new area be planned for 'one class' only. The argument for the view that it should be one class is that people may prefer to live with people of their own class. The opposite view is that one-class housing estates had certain definite disadvantages. They were 'amorphous' socially, inert and lacking in leaders and a community focus. These criticisms were levelled not merely at housing estates as such but also at 'commuter' districts, 'dormitory suburbs' and so on. It was also held that if people lived solely with people of their own social class untrue and undesirable stereotypes of people of other social classes would be created. This would make for harmful social tensions (Willmott, 1963).

The proposed solution was to create a 'balanced' community by 'mixing' people of different social classes. This would foster leaders, establish social foci and promote a 'community spirit'. Many of the 'balanced community' advocates believed that the absence of a middle class appeared to be tied in with the lack of clubs and organizations. F. J. Osborn who discusses the characteristics of Letchworth and Welwyn Garden Cities (pioneer new towns in England) mentions the strong desire of the planners to minimize the segregation of the classes but admits that people of like taste and habits of life (which often, though not always, means people of approximately equal levels of income) tend to group themselves together. Even in these carefully planned towns, areas could lose 'social tone' (Osborn, 1946). It does appear that the middle-class advocates of the mixing did believe that clubs and organizations were necessary.

However, mixing in itself does not always lead to balance. Woodford (in Willmott and Young, 1961) showed, in comparison with Dagenham, a certain amount of tension. Willmott has suggested that neighbourhoods might be planned as one class but within mixed communities. The length of time of residence seems to be another important factor affecting attitudes to other people. Somewhat similar tensions were noted in high flats in Pimlico, London (R. Glass *et al.*, 1964). Gans has pointed out that homogeneity of residents turns out to be more important as a determinant of sociability than proximity. He contrasts, in the USA, the apartment house located in a transient heterogeneous neighbourhood with an apartment house located in a neighbourhood occupied by a single ethnic group. The former is a lonely anonymous building, the latter a bustling micro-society (Gans, in Pahl, 1968).

Mobility in urban societies

Geographical mobility In the last ten years or more, there has been a considerable movement of population within a number of countries. There is nothing new in this; there have been great migrations in history, but the volume and extent of movement in the last century have been very considerable. Modern transport and communications, in spite of certain drawbacks, have helped people to move.

However, the word mobility will carry meanings which must vary according to the wealth and social position of the people concerned. A millionaire may regard a supersonic flight across the Atlantic as a frequent journey; an old age pensioner may find it impossible to afford a short bus journey to keep in touch with relatives or friends or indeed may find the short walk to the nearest shop a physical problem.

The actual movement of a family may affect kinship relationships. Bott suggests that the network of a family will become more loose-knit if the family or other members of the network move away either in a physical or social sense. Firth, however, has gone so far as to state that because there is a marked physical separation from the kin of a household, it may actually increase the possibility of selecting relationships and that this is one of the ways in which urban influences towards individuation and impersonal isolation are counteracted (Halmos, 1964). A certain amount of research, not entirely conclusive, has been done into the different kinds of visiting patterns of relatives, often based on frequency of visitation as a measure. Townsend (1963) thinks that 'although much close investigation of middle-class life ... remains to be done, the evidence that has so far emerged suggests only small variations in function and frequency from working-class family life.' Bell (1968), on the other hand, thinks that measures of contact are very dubious for families that are geographically scattered. He also thinks that possibly the working class, in the area of his study, had a geographical base, which was not found in a middle-class area. The working class depended partly on proximity which gave rise to face-to-face familiarity and common shared experiences whereas the middle-class sense of community is based more on shared opinions and values which can be disseminated as easily by reading matter and other forms of communication as by common residence (Robson, 1969).

Social mobility There are a number of ways of considering social mobility. We can examine the changes which may occur as between two generations—if a son moves into a higher social position than his father—'inter-generational' mobility or, if an individual moves up in his own career—'intra-generational' mobility. Here a distinction may be drawn between salaried employees who occupy positions that are parts of a bureaucratic hierarchy and those that are not. According to this distinction, middle-class people would have a greater potential for intra-generational mobility than members of the working class (Dahrendorf, 1959). Watson thinks that most middle-class people enter into careers of an almost identical pattern and names as 'spiralism' their progressive ascent through a series of higher positions. This ascent is linked up with a residential mobility (Gluckman, 1964), not necessarily with bad consequences. In Crestwood Heights, no matter how mobile each person was, there was a niche for him in the social structure (Seeley *et al.*, 1956).

Urbanization and social mobility In the USA, where a study of considerable coverage was made of a number of urban centres, the origins of social mobility were attributed to a number of causes. As the urban centres grew, migrants were required to fill new positions or to replace urbanites who characteristically failed to reproduce themselves. This process of migration allowed local urban sons of workers to move up into higher positions, e.g., into the middle class, while the vacant jobs were filled by migrants from rural areas. The Oakland data appeared to show that the larger the community of 'orientation' in which a man spent his teens, the higher the status of the job that he held was likely to be. Among men of working-class parentage, size of the community of 'orientation' was directly related to their opportunities for *upward* mobility. This mobility cycle appeared to be only really found in large towns (Lipset and Bendix, 1959). Schnore (1965) criticizes this American study on the grounds that it focuses far too much on the inter-generational crossing of the 'manual-non-manual' line and fails to resolve the problem of rural to urban occupational mobility, the significance of which varies from country to country. One problem that is suggested but not answered in this study would be the effects of the stoppage of migration on the actual social mobility within large cities. It is only an assumption that large cities will always be fed with migrants from rural areas. One can also see the problems facing a country like South Africa. If urbanization and industrialization produce upward mobility what will become of the policy of racial inequality?

The theory of embourgeoisement Goldthorpe and others

have studied the widely accepted thesis of working-class embourgeoisement; the thesis that as manual workers and their families achieve relatively higher incomes and living standards they assume a way of life which is more characteristically middle-class and become, in fact, progressively assimilated into middle-class society. Their conclusion is that the idea of appreciable numbers of manual workers and their wives 'turning middle class' in the way that has been frequently suggested is a highly questionable one and that we must be sceptical of the reliance that is placed on rising affluence, advances in technical organization in industry and changing patterns of urban residence as forces likely in themselves to bring about a radical re-structuring of the stratification of hierarchy; social stratification is ultimately a matter of sanctioned social relationships (Goldthorpe *et al.*, 1968-9).

Berger, who has a similar approach to the study of embourgeoisement, studies the movement of automobile workers to a suburb, a semi-rural community. His initial assumption was that working-class families would learn middle-class behaviour as a result of the 'suburbanization' process. Both Berger (1960) and Dobriner (1963) are interested in the question 'Do the suburbs have a sociological reality?' as opposed to an ecological reality. Dobriner, like Berger, points out the familiar suburban features, homogeneity, commuting pattern, geographical mobility, economic dependency on the central city, familism, the middle class, informality, conformity, excessive participation and so on. The conformity, 'the involvement of everyone in everyone else's life' is supposed to make suburbia classless or one class, but Berger has rightly pointed out that previous studies of suburbs such as Levittown, Park Forest and Crestwood Heights (Gans, 1967; Whyte, 1957; Seeley *et al.*, 1956), were selective and were of middle-class suburbs. There is no reason, however, to believe that residence in a new 'tract' suburb should generate a uniquely new middle-class style of life. This is a different view from

Whyte (1957): 'They (the new suburbs) have become a second great melting pot ... as newcomers to the middle-class enter suburbia, they must discard old values.'

Berger and Dobriner stress not the ecological residence in an area as an explanation of suburbia, but the centrality of class as a variable in explaining suburban behaviour.

It seems somewhat surprising that in the USA, where one would expect blurring of class lines due to the lack of familiar (European) guideposts to social class and class-consciousness, there is a tacit recognition of the patterning of class differences. This may be because *some* suburbs have made the qualities of the middle-class more visible (Dobriner, 1963).

The complex web of urban living

Social organization and social patterns in urban environments

The understanding of social organization in cities seems to have come rather late in the development of sociology. Urban society differs in a number of ways from rural society and social philosophy, even as far back as Rousseau's 'noble savage', had come to look on rural living as the most moral or 'normal' state of human life. Hence, there was a tendency to decry anything in the city, to label it as inferior to the country and to say that it was lacking in certain fundamental things. In contrast, matters rural were called 'good', matters urban were abnormal and bad.

In the total social life of American cities where science, learning, art and aesthetics were initially comparatively insignificant forces, what was good was developed either by recent frontiersmen fresh from their struggle with the virgin forests or by European peasant villagers. The choral societies of Rochester and Milwaukee, for example, had special characteristics derived in part from the ethnic composition of the city's population (McKelvey, 1969). Under these conditions, early observations concerning social organization tended to emphasize either disorganization or lack of organization and, in fact, there was even a 'theory of social disorganization' which fed on the apparent evidences of disorganization supplied by such urban areas as the slums. We cannot lay the blame fairly on the shoulders of the Chicago school; these tendencies were partly a reaction which came from those researchers who 'discovered' social organization in cities.

The urban community, it has now been well established, is a highly intricate social, ecological and economic organization. Its social structure includes all of the simpler forms of social organization, such as primary groups, found in rural societies and it also has some distinctly new forms. Although urban research shows deviant behaviour and social problems which exhibit a divergency of norms and considerable differences between urban and rural life, there is now a continuous recognition that social processes are at work in urban areas to establish social control through assimilation into new forms of community organization. It is these processes which are of great interest which we shall study when we look at what are called deviant behaviours or deviant areas.

Moreover, even quite early in urban studies it was recognized that the city plays an important role in fostering social movements of all types (Burgess and Bogue, 1964) such as the organization of welfare activities.

Complex organizations constitute one of the most important elements which make up the social web of modern societies. Most citizens of modern societies may be born in a hospital, educated in a school, work in one organization or another and to the degree that they participate in religious and political activities they also frequently take part in complex organizations; members of modern societies obtain a large part of the material, social and cultural satisfactions from large-scale organizations. The way to the understanding of modern man and the society in which he lives leads, therefore, to the study of complex organizations. By organizations, we mean social units which are predominantly oriented to the attainment of specific goals such as those which have 'bureaucratic characteristics' (Weber's definition).

It will be found that one of the most neglected fields of

organizational research is that of the comparative study of organizations especially in the same culture and society. Most studies of organizations are case studies of a single organization, e.g. a hospital or a firm.

It will be realized that the study of organizations is of considerable importance. The higher levels of large-scale hierarchical organizations are usually to be found in larger towns. The evolution of organizational patterns in modern society has been accompanied by a number of important social effects. For example, organizational progress has made it possible to be more tolerant of the personal needs and idiosyncrasies of individual members of a society; the needed results can be obtained from them without having to control their behaviour as narrowly as before (i.e. in a society where organizational progress has not occurred). Together with this development there has been an increase in cultural sophistication which has increased the individual's value for accommodation and, at the same time, as a consequence, his possibilities of independence (Etzioni, 1970). Some authors do not quite agree with this view at least as applied to their concepts of urban living. Whyte's (1957) 'Organization Man' is shown by him as a conformist. Crozier thinks that this view comes because Whyte focuses only on the top, successful individuals and that, in fact, the social values of today's organization man are conducive to better possibilities of adjusting in a more independent way (Etzioni, 1970).

In contrast to the fear of 'disorganization' we may wonder whether urban living at times can be over-organized. An example which we can give comes from a small American town. In Sinclair Lewis's *Main Street*—which may or may not be typical—there is a group of Norwegian immigrants and their children. For a considerable time they meet together communally to dance their folk dances and sing their old Norwegian songs. The older folk chat in Norwegian in the sociable *Kaffetur* while men and women delight in wearing the attractive folk costumes.

The younger generation begins to go to the town and discovers the social life which centres round the drugstore—an American focus; they mix with young people other than Norwegians. Other ways of living exert their pressure. The folk costumes are folded away in the carved oak chests, the folk dancing becomes less and less, the singing sounds strange and in the end is forgotten. The social organization of this town has overwhelmed them.

It could be said that such an end would not be found in a large city. A counter-organization of Norwegian immigrants could be formed and there might be enough to withstand the peculiar pressures. In Africa, fears about organizations have been quite differently expressed. It was sometimes thought that the process of urbanization was a movement from the ordered ways of tribal life to the anarchy of the urban jungle, hence the use of such terms as 'disorganization' or 'malaise' applied to the urban situation. In fact, the African town is not a jungle. It has its own specific, highly complex form of organization into which the African who comes to town has to fit himself. We have to view the town as a single field of social relations in which different and sometimes opposed sections of the urban population are closely linked by ties of interdependence (Gluckman, 1964).

Voluntary associations

Voluntary associations are those more or less formally organized groups whose membership is by choice. Social clubs, interest groups, professional associations and the like are well-known examples (Goldhammer, in Burgess and Bogue, 1964). In some countries, voluntary associations sometimes appear to have played an outstandingly active role. De Tocqueville, speaking of political associations in the United States in 1831, writes*

* A. de Tocqueville, *Democracy in America* (1832), ed. R. D. Haffner, New American Library, New York, 1956.

> In no country in the world has the principle of association been more successfully used or applied to a greater multitude of objects than in America. Besides the permanent associations which are established by law ... a vast number of others are formed and maintained by the agency of private individuals.

The voluntary association flourishes in a social setting in which the community can no longer function as an all-inclusive social group. When a community is small and each member is known and accessible to the other and when racial, national, religious and economic differences are absent, the voluntary association can have but few functions. Warmth and friendship are provided by the family, kinship group or neighbourhood. This aspect is not always completely understood by those who wish to 'organize the unaffiliated'. I remember that in one refugee camp in the Middle East some young, American social workers organized a youth centre in an Army hut well-equipped with tables, chairs and games of different kinds. Then they took the young people—all from small communities—on a tour of inspection and asked them what they thought of it. The young people were very bewildered and they did not understand. 'It looks very nice', they answered without much enthusiasm 'but how much shall we be *paid* for attending it nightly?'

In the city the situation may be otherwise. The forms of social differentiation are not only magnified a hundredfold but the sheer numerical effect of large urban aggregates is alone sufficient to necessitate social formations of a much smaller size '*A travers le chaos des vivantes cités*' (Baudelaire).* It was once thought that voluntary associations may act as *communication* belts between different sectors of the population and governmental agencies that are removed from them (Goldhammer, in Burgess and Bogue, 1964). It will be appreciated that the lack of these com-

* *Les fleurs du mal*, 'Les Petites Vieilles: iv' Edition Bernard, London, 1942.

munication belts under certain circumstances can produce a kind of impasse, a lack of communication which may have serious effects. However, the view which was for many years a commonplace in urban sociology, that associations have an overriding importance in the social organization of towns has come to be increasingly challenged (T. Bottomore, in D. Glass, 1954).

One explanation that is sometimes given for the function of voluntary associations is that as the family has lost a number of its protective, recreational and other functions, the voluntary association provides the warmth and friendship that were provided by the family, kinship groups and neighbourhood. I do not think that we should entirely think of the voluntary association as acting as a substitute but consider that it may generate its own specialized forms of warmth and friendship and, in fact, provide a specialized outlet for some kinds of personality.

Voluntary associations and the migrant An aspect of voluntary associations which has received special attention in African studies is the importance of the voluntary association for the newcomer to an urban area. The *abakhaya* or 'home boy' group is the most important one for one kind of migrant. Within it he lives, eats and probably works and the group has its own loyalty. Some men leave the group, which comes from a neighbourhood or village, others still cling to it and say 'Thank God we are not as those townees are' (Wilson and Mafeje, 1965). In west Africa, Little thinks that the voluntary associations which he studied are peculiarly the result of migration, representing the newly-arrived migrant's response to urban conditions. He has been used, in his rural home, to a highly personal set of relationships and so to organize similar practices of mutuality is, for him, a spontaneous adjustment to his environment: instead of weakening tribal consciousness life in the new urban environment tends in some respects to make it stronger. Voluntary associations assist adaptation to the larger variety of new social groupings and network of urban industrial society by providing a new basis for social organization which is all the more important because of the lack of integration in the industrial town in comparison with the rural areas (Little, 1965). V. Pons (1969) has sketched a picture of the newcomer to a central African town, 'a man completely overawed by the strange situation in which he found himself, frightened by daily life in the town.' These views have not passed without challenge. Some writers have pointed out the danger in applying the word 'tribal' to some African associations. The *compins* of west and east Africa represent a later phase in the process of urbanization; all these various forms have their basis in tribalism but cannot be described as simple manifestations of the tribal principle. This does not, however, necessarily mean the emergence within the towns of a class structure on the Western model. Epstein (in Breese, 1969) disagrees with Little over the migrants and associations. He thinks the majority of immigrants are 'non-joiners'. Banton (1966) has the interesting suggestion that industrial sociology, little studied in Africa, should be used to find out whether informal relationships among workmen are modified by such features as tribalism and kinship.

Social class and voluntary associations One of the more controversial issues in urban sociology is the relationship between voluntary associations and class at least in so far as this has been demonstrated by research. The generally accepted view seems to be that in working-class areas voluntary associations, at any rate those of a more lasting type, attract less membership and are probably less in number than in middle-class areas. This difference in participation according to Robson, can be demonstrated as variations in social behaviour provided that personal social class characteristics or income characteristics are held constant. If neighbourhoods were considered, different neighbourhoods showed very different patterns of participation in formal clubs and societies. In fact, the neighbourhood appeared more important than the personal, social characteristics of the individual himself in defining the individual's patterns of social behaviour (Robson, 1969). M. Stacey has strongly stated the importance of formal associations for the middle class.

> perhaps the single most striking quality of the study of neighbour relations ... [is] that the quantity and quality of these relations changes remarkably from one social class to another. Working class contacts, on the other hand, are much more likely to be informal in nature. Thus, while sociability may be widespread within the middle class areas, it would appear to make much less stringent demands upon the *total* actors involved (Stacey, 1960).

We must remember that short-lived but rapidly created associations have had success on occasion in working-class areas—for instance, the organization of 'Coronation Street Parties' and parties for other special events.

Boskoff, in looking at voluntary associations in the USA has a slightly different approach which resembles Little's. He thinks that it may be possible to infer that voluntary associations are an effective and desired means for lower class adjustment to cities during the first generation of urban residence. As they find it possible to cast off previous cultural traits, urban migrants then enter a period of transition in which organizational activities are focused on occupations and fraternal activities. He finds, however, that it is difficult to generalize on the relationships between social mobility and voluntary associations. While social class differences were most closely related to patterns of participation in voluntary groups, as expected, attitudes to the community in general also seemed to be moderately significant in distinguishing differences in participation. In particular, those who indicated general satisfaction with the size and operation of the community, as well as optimistic attitudes about the future of the community, were somewhat more likely to belong to two or more voluntary associations. He divides voluntary associations into two types. The expressive associations (garden, clubs, etc.) and associations for the initiation of desired changes in local, national or international institutional organizations. As such social action associations achieve some success and momentum, they become the foci of social movements. High status persons in the USA, according to one piece of research, show a greater diversity of associational interests (A. Boskoff, 1970).

The object of Bottomore's study (in D. Glass, 1954) was to see how far the hierarchy, derived from an analysis of the social prestige of occupations, is reflected in the structure of such organizations (as voluntary associations). He sought to find what kinds of people, in terms primarily of their occupations, belong to voluntary organizations and whether people of different occupational status meet in these voluntary organizations on a footing of equality and—a question which is often raised in discussions about voluntary associations—how far is leadership in the organizations distributed among people of different status?

He discovered, for a small town that he studied, that there is a correlation between occupational status and status in voluntary organizations. The correlation shows some variability, but at the bottom of the hierarchy the members of the working class play a relatively small part in the voluntary organizations (other than their own particular sport, trade union and political organizations) and have practically no share in their government.

There is a connection shown by Stacey and Bottomore between membership of associations and class, but it does seem advisable for us to give greater consideration in future research to changes, continuities and contradictions in participation in voluntary associations and to the factors that help to account for these differences (e.g., the different stages of the family cycle, the nature of geographical mobility as it affects membership of voluntary associations, marital role definitions and occupational experiences). We have also to consider in the study of neighbourhoods the effect that the different stages of residence has on the creation of voluntary associations or on their membership.

Urban social relationships and proximity

As man is not an island we are influenced, sometimes very deeply, by the social patterns that are produced when people live in greater or lesser proximity to each other. By living, we mean not merely the actual residence, the place where people eat, sleep and have recreation, but also the journey to work and the work-place itself. Our examination of these social patterns should commence with the influences that affect individual families under urban living. We should ask how the advantages and difficulties of urban life and its social constraints affect a family within and then extend our study to the various ways in which families of different social categories react to the neighbourhood or locality in which they find themselves. For example, the work of the father (or mother) of a household, the length of the journey to work, which may be determined by their situation in the city, and other urban factors will have important effects on the rhythm of life of a family (Chombart de Lauwe, 1959-1960).

Another important group of social factors which affect both families and individuals are the buildings in which they live or which surround them, influencing their lives. Michelson's (1970) diagram shows the different ways in which walls, floor or ceiling are shared by people in various types of dwellings. The closeness of contact of human beings poses a number of serious social questions.

> But all he did was spread the room
> Of our enacting out the doom
> Of being in each other's way

And so put off the weary day
When we would have to put our mind
On how to crowd and still be kind

Robert Frost, *In the Clearing*, Holt, Rinehart & Winston, 1962.

In the past, there seemed to be only two choices for the study of social relationships in any particular area. If one adopted a crude form of *architectural determinism* one would say that the construction of houses and their general layout in relation to each other produced certain social relationships. Therefore, houses and streets planned in a particular way will produce specific social patterns. The opposite choice was to declare that it does not matter where people live; their social patterns are unrelated to their environment.

That neither view is completely accepted by urban sociologists is partly due to research. The Westgate study in the USA (Festinger, 1950) examined a special area inhabited by unusual people, mature students at a University. The area had, therefore, a homogeneity which would not be found everywhere and the research workers were interested in the formation of channels of communication which might or might not be formed between people. The physical environment appeared to play a part in this process. People who lived in end houses had fewer friends than did residents of the other houses because they met fewer people in casual contacts. The physical distance of people from each other also served to be an important variable in determining who was friendly with whom as well as the arrangement and design of the houses. Festinger points out, however, that the formation of *groups* is further fostered by the process of meeting new people through present friends. The more cohesive a group, the more friendship ties there are within that group, the greater will be the effect of communication in producing uniformity of attitudes, opinions and behaviour and the stronger will be the resulting group standards.

The Coventry pioneer study can be placed in opposition to Westgate (Kuper, 1953). Here some of the people disliked the possibility of meeting a neighbour or the neighbour's children in a common passage, or being observed going to the outside wc. The closeness of contact does not suggest the formation of friendships of the Westgate type and indeed rather stresses the need for the study of privacy. For some people, such as handicapped or sick people, privacy must not militate against the provision of help in emergency. In high flats, the problem seems to be one of isolation in corridors, and in France, the creation in flats of 'interior streets' has to some extent reproduced the social effects of small, open roads (Chombart de Lauwe, 1959-60). It has been thought that the cul-de-sacs of Dagenham (the 'banjoes') encouraged more friendly relations. It is easier for people to think of 'our street' if it is physically framed than if it has no clear end or beginning. (Willmott, 1963). Both Kuper and Willmott agree that the layout may assist more friendly relations when people are compatible with each other already.

Neighbourliness Both Mogey and Pfeil (Mogey, 1956; Pfeil, in Pahl, 1968) have made surveys of professed and practised 'neighbourliness'—apparent reserve and actual proximity. Mogey divided his two areas into the categories of 'neighbour centred' and 'family centred'. The old Oxford area of St Ebbes was said to be 'neighbour centred'; the street was the traditional play space of the children and people were tolerant of other people's children. The housing estate at Barton showed that people were more inturned. They were 'family centred'. Pfeil notes (in Pahl, 1968) that people in Dortmund who have no relatives make friends with their neighbours more easily than people who have their family living close by. Close proximity is not the only explanation of the need for privacy. Another motive is evident: respect for the sphere of privacy. Other

German writers have considered the importance of neighbours. 'They are nearly as much respected as relatives' (Wurzbacher, 1954).

Neighbourliness and crisis Even if we thought that neighbourliness must usually be measured by the kind of greeting given or the length of time spent chatting to a neighbour, there are occasions of crisis which transcend the ordinary meeting points of people living in proximity. Sudden illness where help is urgently required and the crisis of death are outstanding examples. I once asked a young Welsh woman in a small town to whom she would turn for advice and help in a really serious emergency such as death. She answered 'To the neighbours, of course. This is so in our locality at such times. But you must not think that this happens all over our town. There are localities where this is not done.' In the study of 'Avenue 21', a street in a Congo town, people who lived in a relatively self-contained and self-sufficient neighbourhood, who had learnt to differentiate between different tribal elements, showed rapid and efficient reaction when one of their area was killed in a street accident. They contributed in large numbers to the funeral ceremonies (Pons, 1969).

Thomas Chalmers over 140 years ago believed in the value of neighbourhood feeling. He made an appeal for welfare matters to be left to the neighbourhood 'The report of a perishing household would inflict a discomfort upon the whole neighbourhood and call out succour, in frequent and timely forthgoings, from the contiguous families ... the local system, in short, would bring a sense of character into more quick and habitual play among the intimacies of a city population.' It seems that Chalmers may have been thinking of a somewhat more homogeneous kind of town population than is found today (Chalmers, 1821-6).

There are, in fact, certain reasons why the study of neighbourhoods has attracted the attention of welfare workers and town planners. There is a view that cities are facing a crisis in human relationships. Doxiadis (in Verney, 1969) says, 'We cannot make an absolute choice between local neighbourhood and the city wide community. We need a system of communities from the smallest to the largest.' However, Schorr, discussing slums, says further, 'The block dweller does not feel at home outside his neighbourhood ... he has little interest then in ... the pattern of the city space beyond his neighbourhood ... neighbourhood should have a degree of face-to-face contact among its members.' He thinks that this outlook among some city dwellers shows that the inhabitants of our cities are losing touch with their neighbours (1964). Gans suggests (in Schorr, 1964) that buildings provided in cities should be varied. The site plan should contain a variety of house-to-house relationships, so that residents who desire a large number of social and visual contacts and those who prefer relative isolation can be satisfied.

The neighbourhood unit The idea of creating a complete neighbourhood by town and site planning seems to have been first associated with Clarence Perry (about 1923 in Mann, 1965) in the USA. Perry thought of a locally planned residential area with a common architectural pattern and location of such amenities as schools, playgrounds, shops, parks, play spaces and so on. Although this was before the era of the very great traffic problems, Perry had a road pattern designed for pedestrian safety as well as vehicular access which we might now call the Radburn plan, as it was intended both to separate pedestrians and motor cars and beautify the surroundings. The plan was to be 'one class' socially and it was hoped that the residents would be drawn together through the use of common recreational facilities. Perry seems not, however, to have over-emphasized the communal side of his plan and he showed a very commonsense view in his idea that important amenities like schools, shops and so on should be placed in the right position. Unfortunately, his ideas received very little practical application in many of the council or private estates built in Britain even beyond the end of the Second World War.

Recent developments have shown some improvement on this position and the newer idea of 'neighbourhood unit' that has become acceptable to architects consists of an area containing roughly the population that will support one primary school (this population varies between 5,000 and 8,000 or so). The primary school must be placed in such a position that children and mothers do not have very far to walk to it. Shopping facilities, medical and other services and certain other amenities such as a community centre, churches and so on, will have to be provided with the same attention to distance, so that the elderly person as well as the child has only a certain length of road to walk. The problem of fitting all the houses in without overcrowding, as well as making the correct adjustments for distance, is quite a large one for the enthusiastic planner.

For the urban sociologist, the problem is different. The study of the neighbourhood is valuable because it represents certain social phases in the lives of human beings. The baby and young child have their initial social contacts—but important ones—limited in their geographical extent by the mobility both of the child and its mother. The home and nearby homes, may be the first steps in the world. There is then an important extension when the child walks to the primary school a few hundred yards away, on the neighbourhood unit. The journey to school and the social contacts made there are equally important. With increased mobility and higher schooling the social contacts begin to widen and this process becomes more marked with work. There may be then an important break in the individual's life with new work, involving leaving the neighbourhood, or on marriage which also could mean a departure from the neighbourhood. When a woman marries and has a young child, her immediate neighbourhood is important because her freedom of movement is limited and her contacts for a time may be mainly with other young women in a similar position. The view of some English sociologists is that friendships made during the early stages of this neighbourhood experience are less lasting than friendships made somewhat later. American views, on the other hand, are slightly different. They argue rather more that friendships made during the early stages of this neighbourhood experience, e.g., at the school level, are friendships which last. In old age the position of the individual changes. There is a greater degree of dependence on the neighbourhood; the accessible corner shop within walking distance or wheelchair distance may be necessary not merely for vital food supplies but also for social contacts to mitigate loneliness and isolation. The quiet public gardens and the secluded seat may be places of rest and contemplation.

We must remember that neighbourhoods may undergo considerable changes in population age-ranges over a period of years. One estate in Birmingham, thirty-five years ago, was characterized by a very large number of young, married couples with children aged one month to ten years old. These children, who used to cram the very active Sunday School and swarm over the children's playing ground, have married and left the estate. The estate which used to have a few old people now consists of ageing couples.

There is a belief which is found lurking behind the physical concept of the 'neighbourhood unit' that people will become more 'neighbourly' if they live in a neighbourhood unit. Sometimes the analogy with the village is put forward to show that people living in a small community are neighbourly. Sprott points out that the village is practically always seen through the eyes of the middle class (Sprott, 1958). The working class may have regarded the village quite differently as we find in such books as Flora Thompson's *Lark Rise to Candleford*. Moreover, the size of the neighbourhood unit is against any real comparison with the village. Even if we accept that there is something neighbourly about 300 to 400 people living in a village—which is not necessarily the case—we may be unable to say the same of 5,000 people. The really important thing is to find out what kind of social relationship exists between people rather than to think about the geographical area in which they live. It is possible that social control is probably stronger in small residential groups than in the larger neighbourhood and that small residential groups, where they still exist, may react more readily in a communal fashion than areas where dependence on the neighbourhood has declined. Toby (1964) has an amusing and almost incredible description of 'The Great Bank Robbery', an incident in a small American town. Such was the social ethos of this little community that the thieves never had a chance. Strangers were immediately noticed, observed, subjected to gossip and, where necessary, the community could be aroused against them.

The concept of community in urban studies

We must now analyse one of the more puzzling ideas which has appeared in sociology. The word 'community' has been given no less than ninety-four different definitions in sociology (Hillery, 1955) and in ordinary speech there are enough ways of using this word to create some confusion. For example, we hear the word 'community' used in such an expression as 'community spirit' in which sense it probably implies something that is considered useful, desirable or necessary in a village, a football team or a school. It can also be used of a collection of people with some purpose, such as the Community of the Resurrection—a body of men living in a particular religious order—in which instance there is a strong element of unity or oneness.

German sociologists have an advantage in their own

language for they are able to choose from two words to express the idea of 'community'. The word *Gemeinschaft* is meant to express a broader concept—the community in general—though in Tönnies (1955) it is rather geared to the expression of a particular kind of social relationship. *Gemeinde* which is also translated as 'community' is considered by R. König (1968) to be a social unit on a local basis in which men work together. Possibly from this definition comes the notion of community as a place where people spend their entire lives though there are some difficulties in adopting this not unpopular definition.

A development of this word *Gemeinde* is twofold. On the one hand it may be considered as a sociological concept by means of which every kind of relationship between structures on a local basis can be understood. From this kind of approach a city is a settlement in which common and daily life shows a tendency to be polarized either in the social aggregates of public or private life. Urban communities are considered to be in opposition to rural systems (Ruegg, 1969).

If these approaches seem over-complicated they do, at least, give two leads. We ought to look at small social units and to look at them on a local basis. Studies of preliterate societies, e.g., a single tribe, often have the advantage that they include all the people living in the same local area and having the same social and religious characteristics, perhaps even the same economic life. Some Western social units, such as certain villages, though in contact with urban influences, may display similar features, if not all. The oft-quoted village of St Denis (Miner, 1939) reveals a social culture with a high degree of internal social integration. The *habitants* live in terms of common understandings which are rooted in tradition. Family life is important; there is little disorganization or crime. However, while St Denis appears to be a 'community' it is difficult to say which features are more essential to its existence as a community than others. The sense in which the term 'community' is applied to St Denis is very different from that in which it can be applied to a 'metropolitan community', an 'urban village' or even to a 'town'. There are other forms of community organization in which 'order' and 'stability' can be found. For example, in a middle-class suburb there may be many voluntary associations which, at first sight, appear not to be parts of any one identifiable community but may soon reveal a multiplicity of interconnections. The slum community of *Street Corner Society* (W. F. Whyte, 1942) shows both social organization and social control yet it is considered to be 'disorganized'. St Denis, therefore, ought to be considered as a community, but only one kind of community (Worsley, 1970).

Despite these problems, König argues that the community is a *unit* and that though it has many different forms, it remains an entity in its own right. For him, the community is a basic form and it does not disappear or dissolve in the wake of urban and industrial revolutions as has been often argued or implied (König, 1968). In fact, studies in Germany on the effect of industry and urban influences on village life, suggest that some European rural areas may have been spared the complete destruction of the foundations of a rural life (Wurzbacher, 1954), which can be found in parts of the USA and Britain.

We have seen that Redfield constructed 'ideal types' of which his 'little community' was an example. One of the major criticisms levelled at Redfield's concept of community is that his whole construction appears to be based on the assumption that the 'little community' is an integrated homogeneous whole, free from tensions and somehow more harmonious and unified than a large community. In fact, such an analysis as the study of Glynceiriog shows that within a small village there were diversions and that the community was ill-defined. Here the two sets of people who comprised the village could not agree to accept a bandmaster from either set, therefore one was sent for from another area who would be acceptable to both groups.*

The study of 'communities' in towns A perceptive and valuable book (Stein, 1960) reviews certain outstanding studies on communities in the USA. Stein's book looks at Warner, Park, Lynd and Seeley. He is concerned to find a theory of community, an approach to community which should be able to account for the main patterns of change found in each of the communities examined. He doubts whether the study of any particular American community will be validly applicable for all American communities. His view of the Chicago studies was that they contained special features which had to be clearly identified before they were considered relevant. The ghetto of today differs considerably from Wirth's ghetto. He found in the two Middletown studies in contrast to Chicago that change arose from different processes. There was, for instance, a great flow of immigrants to Chicago. Stein's study is valuable because he is moving away from the rural–urban contrast model and is concerned with social change in the city and produced by the city. He thinks that each community study must describe the social structure of the community as it has changed and is changing over a period of time.

The chief criticism of Stein is the title of his book. Was he really describing the 'eclipse of community' and did he possess an unjustifiable nostalgia for the past? He does discuss the processes of urbanization and bureaucratization in rather pessimistic terms. In looking at the 'final breakdown' (in the *Yankee City* of W. Lloyd Warner), he says, 'relations between [workers and managers] ... were henceforth to be mediated by impersonal contacts ... their sense of belonging to a common enterprise, or to the city itself, seriously declined.' This quotation typifies the problem that Stein was facing. He is clearly observing a change, but he tends to have a subjective rather than

* R. Frankenberg, *Village on the Border*, Cohen & West, 1957.

objective approach. This is, perhaps, evident in his examination of the sense of belonging, the feeling of change which some inhabitants have and the feeling of some that they are becoming strangers in their own town. It is unwise to divide townspeople into two divisions, strangers and townspeople, and to assume that this division represents the decline of community. Community life does exist in cities, sometimes strongly, and it is not a rural type of community. In examining the readjustment of immigrants, Plotnicov (1967) says: 'Aside from individual adjustment, the developed and developing social institutions supply organisation, viability, durability, a measure of social security and the satisfactions that allow urban life to be worthwhile to its inhabitants.'

The empirical view of community: community research and neighbourhood studies The attitude of British sociologists to community is a practical one. Mann (1965) suggests that the word 'community' is valueless as a sociological concept unless it is defined in terms of observable behaviour—an operational definition. We have a long way to go before an operational definition is reached.

The Berinsfield study is in the tradition of community research. The aim was to ascertain what value the word 'community' had in application to the particular group of people concerned. The territorial unit is the starting point and the social relationships which exist within this unit are investigated. Four types of community relationships are considered. Common bonds through shared interests; networks of interlocking memberships in voluntary associations; common membership of reference groups; and the feeling of living in a friendly, cohesive, territorial group.

A 'phase hypothesis' is postulated in the development of social relations on this new estate. The second phase appears when the estate is so big that it is no longer possible for even the more active to feel that they know each other. It would be interesting to know whether actual historical processes parallel this development.

The continual influx of new members threatens the intimacy and the nascent small group structure disintegrates. Families tend to withdraw from largely *ad hoc* relationships.

The common bonds were largely created because most people were new tenants and the important role of children was that they produce either co-operation or conflict.

The four aspects of community were by no means identical in the development over a period of time. Morris distinguishes three main dimensions of community: the type of social group; the type of physical grouping; and the extent to which social and physical grouping coincide. 'Community' describes the case where primary social relationships are coupled closely with a small, physical group. 'Association' describes the case where social ties are secondary and there is not correspondence to a physical group. This distinction seems to be of considerable practical importance. Most housing estate studies find cleavages developing which may be encouraged by the activities of the community centre. It could be argued that higher correlations between the different aspects of community would have been obtained if social change had been less rapid and the social structure more cohesive. As families became more settled and the social structure more cohesive, one might have obtained higher correlation between the various aspects of community. 'In Berinsfield, differentiation developed more rapidly than cohesion. Common bonds tended to decrease in importance' (Morris and Mogey, 1965). A somewhat similar situation is discussed by Ruth Durant (in Pahl, 1968), who points out that a housing estate will never exactly resemble an ordinary town.

It must be remembered that conflict, as well as cohesion, may be of importance in a locality in so far as it may develop a common consciousness of living within the locality (Pahl, 1970). During the Second World War, the inhabitants of a small town were made very conscious of their town when it was 'invaded' by several thousand scientists. What the town meant to them—the original inhabitants—and to the strangers was a different thing and it took some time before people became adjusted to the new residents. In quite small communities, where people are living close together, they may try to avoid clashes. Social control through gossip, for example, must be used carefully and tactfully in an indirect rather than direct way (Elias and Scotson, 1965).

5

Change and other urban phenomena

The newcomers

Where do the migrants come from? The movement of population has sometimes been considered a symptom of basic social change. 'International' migration, like some kind of social upheaval, has had a dramatic appearance. 'To the earth folk, seeing their plots diminish, while their offspring increase, tidings had come of a vast land on another continent where fertile soil was to be had almost for the taking.'*

However, equally important for the people who take part in them are the 'internal migrations', the movement of people from country to town within a land, from town to town, or within a town. One old lady who waited to be moved two or three miles in a new town said, with as much feeling as the Swedish migrants—'You may not think much of this old cottage with the allotment by the hedge and the muddy path to the spring, but I have seen this daily for fifty years. I do not know how I shall be when I leave it.'

Historically, cities may have been dependent upon migration for their growth or even for the maintenance of their population. This was the reason for the description of cities as 'deserts of men'. This view rather implied, too,

* Vilhelm Moberg, *The Emigrants*, Max Reinhardt, 1956.

that people were attracted to cities by their 'pull' or special attraction. The opposite view was that migrants were 'pushed', e.g., by economic stringency out of their own locality to the city. It is sometimes thought that migration to towns by-passes a 'folk stage' (Hauser and Schnore, 1965). Whether we accept or reject these views entirely, it does seem that in the past, industrialization was accompanied by vast shifts from rural to urban areas (Jansen, 1970).

We have still to remember that it is not easy to distinguish rural from urban nor can we consider the process of migration as being something entirely uniform. In some pre-industrial societies the familial and religious organizations in their most developed forms are urban rather than rural. Moreover, there are major differences between mature industrial systems and those that are becoming industrialized (Merton, 1959).

The traditional view that populations are normally static and that a positive push or pull is necessary to overcome inertia is no longer realistic in the conditions of modern urban societies (Jackson, 1969).

Who are the migrants to urban areas? Most writers on migration do describe the selective nature of the process. Not everyone migrates and certain groups are more migratory than others. This is partly because the migration process is educationally and occupationally highly selective. Musgrove (1963) thinks that the 'migration of the elite' in Britain—which has one of the most migratory populations in the world from an internal point of view—is important because in the leadership of voluntary associations, immigrants may take over from the local people. Where this does not happen, local government may be impoverished.

The process of migration varies according to a number of factors. In many Indian villages the process of migration to urban areas is strongly affected by long-established pathways of out-migration and by the custom of alternating between village and town life, like the Howrah blacksmith who alternated between work in a family unit in the city and on his family land. Probably the ancient Indian civilization must not really be regarded as rural, but urban in its traditions. However, migration to urban areas may affect the delicate balance of the village. More men than women may migrate (Turner, 1962) and the lack of old people in a migratory movement may cause instability. In Latin America, the rural–urban movement is considered by Germani to be the expression of an actual mental change in the outlook of people (in Breese, 1969).

This alternating shuttle movement of families has not the same serious social aftermath as the movement of isolated individuals moving into urban settings for which they are unprepared both psychologically and socially. Individuals thus circumstanced may have a sense of rootlessness which could provide a seedbed for all kinds of anomic movements. We have also to differentiate between movements which are planned and controlled by 'welfare state' considerations and those movements initiated spontaneously by individuals.

The settlement of the migrants in urban areas This has been examined in two possible areas of work. First studies of immigrants in Western cities such as in the United States and in Great Britain have concentrated on the reception of the migrant and his 'integration' into the urban situation. We may assume—it is not an all inclusive assumption—that a migrant family might differ from the natives in some respects, e.g., in his social level. The migrant family may have new patterns and may have some elements of disorganization due to the migration process itself.

The process through which a migrant will pass is not unlike the basic process of an individual's socialization except that the migrant starts from an already given social basis. He will then go through a process of personal readjustment and if he is dispersed institutionally, that is, not concentrated in sectors of economic, political or cultural spheres, will cease to have a separate identity. An important role is played by relatives and friends (Eisenstadt, 1970).

Acculturation and assimilation

These are points on a continuum. They can extend from a point of maximum difference between people to a point where sections or members of an immigrant group have become indistinguishable from members of the host society. Acculturation refers to the borrowing or acquisition of certain cultural traits from one society by another. It involves not merely a change of behaviour but even a change in the conception of oneself and it takes place without complete integration into the major social institutions of the host area.

One problem that the migrant faces is that he may be working at the lowest level in the urban industrial society but at a higher level in the scale of occupations for the total society because urban industrial society has relatively few positions comparable to those in the rural or small town society whence he came.

Whence did he come? Sociologists have largely ignored the fact that some knowledge of the social organization of the community of the migrant and also the social organization of the host community to which he comes is crucial to an understanding of the process of assimilation. American sociologists would like to know under what conditions assimilation takes place and under what conditions it is most difficult, e.g., how is the Mississippi Delta negro assimilated into a Northern industrial city? The process may take generations (Schnore and Fagin, 1967). The study of this process of assimilation by observing the social organization of the host community as well as that of the migrants' own community would seem to be an important field of work in Great Britain as well as in the USA. We must point out that coloured immigrants may be discriminated against. Discrimination is not a function of lack of acculturation or lack of a class position. Moreover an industrial society, unlike a traditional society is heterogeneous, stratified and pluralistic. This means that complete 'assimilation', whether cultural or structural, is probably impossible, yet at the same time there will be many groups and associations which the migrant is free to join or ignore (Jackson, 1969). On the other hand, a rigidly stratified system will prove a major obstacle to acculturation (Kobben, in Fröhlich, 1964).

Urban values and migrant values To presume that the migrant has a completely free choice is an error. His total experience in his society of origin may have been such that he finds it easy to accept a limited area of competition and decision making. The organization of society is a crucial determinant of how successful a migrant is in becoming assimilated. Important people are those Fagin calls 'gatekeepers', those key people who stand at the important points in the migrants' path, such as the police, social workers, bartenders, landlords and so on. They have decision-making power that either facilitates integration into the community or completely thwarts it.

Urban values and rural values The second field of study of migrants in urban areas has looked at the possible differences between 'town' and 'country' in such countries as Africa. For the African who comes to town, life might appear to have changed considerably. His job may be divorced from his childhood experiences and the urban life may mean contact with people of different customs and even unknown traditions like the contact of the Ibo student who found, to his surprise, that he was expected to kneel before his future Yoruba father-in-law.* We are dealing, too, not merely with the effect on individual persons of migration but also with the effects on a system. The contact may be a two-way process; the migrant returning to his village may represent a clash between 'urban' systems of values and those of the village (Peristiany, 1965).

Incapsulation (or *encapsulation*)

This is the name applied to the cultural process whereby migrants, though living in the city, are yet not of the city, but retain other characteristics. The 'Red' Xhosa, while spending many years in town, are apparently carefully insulated against the typical diversions that the city has to offer. The highest value here is put on roles either belonging to the country home or derived from the country home. The greatest fear is that of becoming entirely detached from the old network of relationships in the country and encapsulation therefore implies some resistance to urban influences (Mayer, 1961). A somewhat similar pattern is described by Pahl (1970) for Great Britain: if countrymen from surrounding areas move into a town and are surrounded by friends and kin from their area of origin they may stay encapsulated for a hundred years or more. Only when men see the situation in wider terms and become aware of their class position in the wider society can they be said to be truly urbanized—in a sociological sense.

In a number of African towns two systems of relations between town and country may be maintained. Earlier African studies tended to emphasise the migrant's 'tribal' background. He was considered to go through a slow and perhaps painful process known as 'detribalization' until he was 'urbanized'. The difficulty with this view of 'urbanization' is that it assumes there is a change going on in the

* Chukwuemeka Ike, *Toads for Supper*, Collins, 1965.

same way and in the same direction. It also may presume that there is a final stage of 'urbanization' to be reached. Gluckman has forcefully stated an opposite view, not necessarily entirely acceptable, in the phrase: 'An African townsman is a townsman, an African miner is a miner'. When is a man a 'townsman'? Rankin thinks that the notion is essentially a cultural one, tied to a particular local concept. We must beware of descriptions of ways of life which can be applied only to one area or to one tribe. Perhaps Southall and Frankenberg have the right approach when they describe the process by which urban roles become more highly specialized. Their method is to analyse the internal process of a community whether urban or rural (Southall, 1961).

Physical decay and social change

One of the most important aspects of urban life is the way in which the physical appearance of the towns can change. A few towns appear to resist this process, such as the mediaeval Carcassonne, but in many towns buildings decay and are demolished, land use changes and new buildings are constructed. In Britain and in a number of other urban industrial countries, the actual physical decay of buildings may be seen in 'twilight areas' which means essentially housing neither by statutory definition unfit (slum) nor good enough to warrant improvement and next in line after slum clearance property. Sometimes a small event can produce a twilight area. The removal of a bus station can lead, for example, to a change in the direction of pedestrian traffic, to declining shop trade and a general state of decay (Medhurst and Lewis, 1969).

The zone in transition Burgess and others described a zone in transition as part of the concentric zones theory of city development. A basic aspect of this theory was the competition for the use of land, especially in the inner parts of the city. The changing nature of the zone in transition has provoked the description 'a region which is always becoming but never really being' (D. V. Glass, 1935).

The competition in land use, especially for houses, represents a more general competition for the use of available material resources. In countries with a 'welfare sector' in the national economy such areas may induce an interference, limited but significant, with the rights of property, which modifies the 'free for all' outlook that might otherwise prevail. Notwithstanding the allocation of resources in the public sector (e.g., local authority housing), the private house owners may remain important. The zone in transition displays a class struggle over the use of houses and, in fact, this class struggle is the central process of the city as a social unit. Max Weber, for instance, pointed out that class struggles emerge wherever people in a market situation enjoy differential access to property such as domestic property. In 1836, a market gardener in Victoria, Australia, purchased for a small sum a piece of land for a vegetable garden. He thought himself unlucky; the ground was of poor quality. Today, because it is in the centre of the city of Melbourne, the land is worth an enormous sum.

As there are different classes of people in the city, so there will be types of housing, varying from outright private ownership to people who are just tenants of a room in a lodging house. These types of housing will have a definite territorial distribution which depends largely upon the age and size of the buildings in the different zones.

Individuals and families over a period of time will move from one housing situation to another. Their movement will depend upon such things as availability of capital or houses or even selection criteria for public housing. In a number of cities the lowest housing standards are undesirable for health and welfare reasons and the housing may be considered 'illegitimate' by welfare state standards (Rex *et al.*, 1967).

The people who live in the 'inner city' can be divided into different sections. There are 'ethnic villagers' isolated from significant contacts with most city facilities, there are the 'trapped' who cannot afford to move when their neighbourhood is invaded by lower-status immigrants or by non-residential land uses, such as the construction of factories, and the 'downward mobile' such as old people on small pensions. The zone has also had an attraction for 'cosmopolites'—artists, writers, musicians, students and so on (Gans, in Pahl, 1968). In London, Westergaard found a large number of young adult people and a very low proportion of young children. He thinks this is an exceptional specialization, no doubt a permanent characteristic of an inner metropolitan zone, which offers special opportunities for young unmarried people, but fails to find homes for them at a later stage in their careers (in R. Glass *et al.*, 1964).

Within the zone in transition, there will be different immigrant situations. Some migrants may keep contact with home, others will not want to. Discrimination may be encountered though conflict may be modified in a number of ways, e.g., multi-racial clubs and societies. The loss of kin and friends may produce personal demoralization. What is created is a segregated and problem area where the social situation is not stable. Even a decrease in the number of people does not really relieve the pressure because those who want houses are attracted by labour demands (Rex, 1967). Although the zone in transition may contain a number of foreign-born immigrants (Chombart de Lauwe, 1965), by no means all immigrants settle initially in the zone and subsequently move 'up and out'. Cities which experience high rates of in and out migration, especially high-status migrants from other urban areas, may have immigrants moving directly into the outer city and the suburban city (Jackson, 1969). One could visualize a situation where there is no zone of transition at least for external immigrants.

The ghetto This kind of segregated ethnic community is familiar from European examples and the Chicago ghetto of Wirth. The Negro ghetto has received recent attention for the socialization patterns it displays and the picture of broken, unstable families, feckless attitudes towards family planning and non-compliance with average 'American standards' of morality and living presented to people *outside*. We can only understand the problems of the ghetto dweller in a Negro ghetto if we realize certain underlying concepts. He suffers from deprivation of the good things of life and from exclusion, which means that he is fettered to the ghetto. To be mobile and leave the ghetto means having both a stable family life and a good economic position. One ghetto Negro father complained of his low family status because he was not the main source of family income. This illustrates Parsons' (1949) position on family status which asserts that a man's status in the family is mainly dependent upon his status in the community. 'You must have a job before we give you food' said his family to a young, unemployed southern Italian. 'Without a job, you are nothing.'

The economic handicaps of the ghetto make it very difficult for a man to provide for his wife and family within the structure of life-long monogamous marriage. It also makes it difficult to inculcate a conception of morality that will teach the children how to cope with the social situation of the ghetto. This picture of a Negro ghetto shows how different a ghetto may appear today from that outlined by Wirth (Schulz, 1969).

The slum There are two aspects to the commonly used description 'the slum'. The first aspect relates to a district which is in 'an extreme state of structural or environmental decay brought about by economic, social or other factors, where the standard of fitness for purpose has fallen below the level at which palliatives might be acceptable.' This definition can be applied not only to housing but to schools and decay can be produced in several ways (Medhurst and Lewis, 1969). The second aspect, perhaps the most difficult part of this definition, relates to the people who live in slums. Who are they? Do they differ from everybody else? Is the slum, as Glass would put it, 'a concept' as well as a concrete fact? (D. V. Glass, 1935.)

The social welfare approach to slums was and still is conditioned by the possibility that certain social problems, e.g., crime, ill-health and immorality were linked with the people of the slums. Some middle-class rural people were greatly surprised when they received 'evacuee slum dwellers' in the Second World War. One observer comments acidly on this situation. 'The conditions of extreme poverty are generally quite unknown to the ordinary citizen' (*Our Towns*, 2nd ed., O.U.P., 1944).

Elizabeth Gittus distinguished between 'residual areas' —dreary or obsolescent dwellings of Victorian tenements —and the 'transitional area' where there has been a change in the occupancy of the buildings and in the essential character of their use. As a student, I lived in a delightful old Victorian house where three students had rooms. A London friend said, 'I could make this into a really good slum dwelling by sub-dividing it still further, letting every room or even parts of a room at high rent to people hungry for accommodation.' Gittus also stresses the changing character of these areas. The residual areas were comparatively respectable, the transitional areas contained the deviants but even so, the social problems only applied to the minority of the population (Medhurst and Lewis, 1969).

The relationship between slum dwellings and social life

It is difficult to ascertain whether there is a true causal relationship between housing and its effects on human beings. However this may be, in one direction the evidence seems to be overwhelming that extremely poor housing conditions do influence social behaviour and attitudes. This situation is one of desperation beyond human endurance:

> I am not hungry for berries
> I am not hungry for bread
> But hungry hungry for a house
> Where at night a man in bed
> May never hear the plaster stir as if in pain
> May never hear the roaches
> Falling like fat rain
> Where never wife and children need
> Go blinking through the gloom.
> Where every room of many rooms
> Will be full of room ...

('The Ballad of Rudolph Reed' in the *Beaneaters* by Gwendolen Brooks, 1960, permission of Harper & Row, publishers.)

It will be noticed from this poem that certain elements seem important to the writer. Apart from dilapidation and decay, the overcrowding, which also carries with it a threat to privacy, is mentioned. However, crowding, though an important factor in influencing the slum-dweller, must be considered with other deprivations. McGee points out that the Chinese in some Asian cities, though both over-crowded and lacking in privacy, retained strong cultural features, showing neither strong disorganization, high crime rates or immorality (1967). There are other features, too, which are also apparent. If we call a house a slum, this may be to tell the tenant that he is a slum dweller, and living in poor housing may itself influence motivation and self-evaluation. 'I do hate to be "looked at"' said one protesting old lady in a London slum. We must remember that there is another factor which plays a part in people's attitudes to their lives in the slum. In the past, cities were probably better designed for 'escape' within the city. They were usually smaller and it was possible not only to walk to work but to find leisure time satisfactions and opportunities locally. With the development of motor car transport the poor may have to live where the middle-class lived before the motor car. Barriers of stratification and differentiation by income or class or colour have been extended to deny and exclude; the poor find it more difficult to find an escape.

If the answer to slum dwellings was just to move the slum dwellers to other areas and into more commodious houses with a higher degree of privacy, it would be a comparatively simple matter. But we have a peculiar series of problems if we consider this solution. There may be people who both require and want more space but, if they gain it, may have to yield up a neighbourhood that is both familiar and secure. Perhaps those people who live close to each other in terms of distance, or accessibility, may tend to become friends or form closely knit social units, but this can only be conjectured. We do not yet know either, from studies of 'young communities' whether the social influence of physical proximity remains significant over a long period of time. We do know that some families gain in morale through more space because the different family members do not get in each other's way. At Berinsfield, there were people who were glad of the move because they wanted to escape from their extended families (Morris and Mogey, 1965).

The slum: organization or disorganization? A new approach to the study of social organization in the slum is offered by Suttles (1968). He takes an area which apparently deviates from the public standards of the wider community. It has a high delinquency rate, numerous unwed mothers and several adolescent gangs, all facts which would tempt one to think that the residents are inadequate and disadvantaged people who suffer from cultural deprivation. However, he discovered that from the inside his area was highly organized according to its own standards, and residents fairly insistent on their own demands, which required both discipline and self-restraint in the same way as do the moral dictates of the wider community. Conventional norms were by no means rejected but differently emphasized or suspended for established reasons. He does not think that disorganization or value rejection are correct terms to apply to this situation; the residents are, in fact, bent on ordering local relations where the beliefs and evaluations of the wider society do not provide adequate guidelines for conduct. He names this situation 'ordered segmentation' from the larger society. The slum community is in fact a provincial society of its own. This is not an inversion of the wider society.

The measure of such a society is sometimes given as the degree to which it complies or does not comply with public morality. It is possible, however, that other communities support public morality not because they feel strongly that public morality expresses their own view of ethics, but because they feel that public morality provides a set of defensive guarantees—a means of protecting themselves against threats from other groups. People may, however, within the privacy of their own group, use different standards from those that they profess. They may 'vote dry and drink wet'. Moreover, urban growth itself has historically tended to congregate different groups together. These different groups may have different moralities of their own.

The creation of social relations within the slum locality

The most important reason for centring interest in a locality is that the people there simply cannot ignore each other. People who routinely occupy the same place must either develop a moral order that includes all those present or fall into conflict. The initial stage in the development of a segregated slum area is when people who feel that American standards of morality are being compromised simply retreat and abandon parts of the inner city. This tends to aggregate those who are poor, unsuccessful and disreputable in the same slum neighbourhoods, a process which usually precedes the creation of a common social framework for assuring orderly relations.

It is interesting to note that such an area may resemble a prison community. In order to escape complete anarchy, Hobbes's 'war of all against all', some method must be used to reveal which people can or cannot be trusted. While some apprehensions about other people can be assuaged by close enquiries into each other's personal character, the important feature of social relations is the way in which they are tied to locality. It is suggested that this develops as a movement outwards, the women and children remaining close to the house, while the males move out from the house. The *location* of other people becomes of supreme importance. This is because residential unity implies social collaboration. There is an assumption that most people 'belong somewhere' except in certain areas where there is a kind of social vacuum or no man's land (industrial areas, non-residential areas such as roads, even public services areas). These areas are distinctly dangerous because there are no people to take a heartfelt interest in their safety. Only people who really live in an area are expected to take an interest in its safety. They must be residents there. The amnesty is not considered to extend to outsiders and the boundaries are strictly defined—'stay by the house like I told you'.

As transients are often found among slum residents the anonymous present a problem. Safety is felt to lie in knowing other people whose behaviour can be anticipated. This is probably one of the reasons why street corner gangs can be formed as the more obvious social groupings. What is so fascinating about Suttles's work (1968) is his conclusion that the sub-culture of the slum he studied was in fact a means for discovering the moral order and providing a decent world in which its people could live.

The barrios, *bustees, shanty towns*

These various names have been applied to makeshift types of housing which have developed on the outskirts of cities in Latin America, parts of Asia and some other places. Usually they are not in the cities proper but grow on the outskirts. Sometimes, a little land is available for food-growing. In South America, they can represent a large proportion of a national population.

The *barrios* have given rise to two opposed hypotheses. The first is that they are simply slums, blighted areas, and crime and social disorganization are common within them. The second theory is that *barrios* may be semi-rural enclaves which make available new possibilities for urban reconstruction on the basis of neighbourhood communities, mutual aid associations and so on. The proportion of activities which demand active rather than passive participation is much higher in the *barrios* than in 'developed' urban societies.

The social creation of the *barrios* is not simply the result of an organized surge of people into land on the edge of the city, but can be effected through a carefully planned invasion by a group of families who know about urban life and take adequate action to secure permanent dwelling space on undeveloped land which is not provided with services. In some places such as Lima, the *barriada* people appear to contradict the common view of the *barrios*. The people have resided in the urban area for nine years on average and are educationally removed from rural culture. Their families are relatively stable and delinquency is rare, though common in the slums. They are politically conservative. Some houses are even brick and cement built—they are scarcely shanties. The solution to the *barriada*, to evict and relocate the people, is too expensive, or too disturbing politically, or would not meet the needs of the *barriada* people.

The *barrios* pose a number of difficult problems. With

very few exceptions no modern state has allowed large numbers of people to work out their own definition of problems and their own solutions; it may be structurally impossible for a state to support what is virtually conflict against itself. The social planner faces a peculiar dilemma as well. These 'marginal' urban groups gather momentum under adversity and recede before benevolence and permissiveness. This makes it hard for them to be socially integrated into the city. It also illustrates the major question in South American urban living—how the process of accommodating 'marginal' populations to the established urban classes should be carried out.

The nature of urban poverty

Research into slums, *barriadas* and the ghetto arouse interest in the study of urban poverty. Probably the city poor of imperial Rome were no better off in health, in food or in dwellings than a poor Indonesian townsman today, but the technique of communication now is such that many people know that they are poorer than someone else. To them poverty might appear to be a way of life in which they are exploited by another group.

One approach to poverty starts from rural living. Foster's hypothesis states that peasants view their total environment as one in which all of the desired things of life are always in short supply. This 'limited good' image is reflected in the conservatism of a peasant society because individual progress threatens community stability. The general result is that the level of aspiration is low compared to middle-class or even lower-class urban dwellers. A natural disaster such as a landslide might, however, make change attractive. The latent powers of peasants would then be manifest and could be utilized. It seems doubtful whether the 'limited good' image is more important in retarding change among peasants than external economic and social conditions (Mangin, in Bloomberg and Schmandt, 1968).

The culture of poverty theory

Lewis (1968) developed his theory from studies made in Mexico and among Puerto Ricans. He omits primitive peoples whose backwardness comes from isolation and underdeveloped technology. The urban migrants whom he studies have a 'culture' distinctive from that of the broader culture around them. It is both an adaptation and a reaction of the poor to their marginal position in a class-stratified, highly individuated, capitalistic society and it represents an attempt to cope with feelings of hopelessness and despair which develop from the realization of the improbability of achieving success in terms of the values and goals of the larger society. Once this culture comes into existence it tends to perpetuate itself from generation to generation because of its effect on the children. By the time slum children are aged six or seven they have usually absorbed the basic values and attitudes of their sub-culture and are not psychologically fitted to take full advantage of changing conditions or increased opportunities which may occur in their lifetime.

Lewis considers that the most likely candidates for the culture of poverty are the people who come from the lower strata of a rapidly changing society and are already partly alienated from it. The lack of effective participation and integration of the poor in the major institutions of the larger society is one of the crucial characteristics of the culture of poverty. So is their low level of organization as compared to most primitive peoples. There is little organization beyond the extended family and kin. There may, however, be some sense of community (e.g., the *compadrazgo* or godparent kin) and a sense of territoriality, which sets off the slum from the rest of the city. Fatalism and a low level of aspiration are listed by Lewis as one of the key traits for the culture of poverty but he agrees that the context where this happens may make a big difference. Moreover, there may be a capacity for spontaneity and adventure which is often blunted in the future-oriented man. A poor man once said to Doxiadis, 'Once you have pulled down our slums, what will become of our poetry and dancing?' (Verney, 1969.)

There are a number of views which oppose or modify Lewis's theory. The *barriadas* themselves can be quoted as showing the potentiality of the urban poor to display behaviour contrary both to the 'culture' theory and the 'limited good' image. The Watts riots in the USA show that changes may take place in an explosive and destructive manner if no other channels appear to be open. Coates and Silburn while allowing that there are some parallels between Lewis's description and slum life in Nottingham, England, point out (1970) that the crucial difficulty is to determine the 'middle-class values' of which the poor are supposed to be aware but do not live by. Within any area there may be 'problem families' and different subcultures. While there was a lower level of expectation among the poor there was, nevertheless, some expectation. It would be very hard to maintain with any assurance that the poor constitute a single sub-cultural entity.

Deviance and urban living

All social groups make rules and attempt to enforce them. These rules define situations and the kinds of behaviour appropriate to them. The person who breaks rules is regarded as a special kind of person who cannot be trusted to live by rules—an 'outsider' (Becker, 1963). The person concerned may regard his judges, however, as outsiders having no competence to judge him. They may even be deviants to him.

Deviance, it should be noted, is not a quality of the act the person commits but rather a consequence of the application by others of rules and sanctions. For example, in some societies a man who commits polygamy is committing a deviant act punishable by sanctions. In other societies the act would not be deviant. We must be careful, therefore, in studying deviance not to assume that there is something inherently deviant about acts that break social rules or that the deviant act occurs because some characteristic of the person committing the deviance makes it either inevitable or necessary that he should do it.

The problem is how to find out which rules are applicable in a modern society. Modern societies are highly differentiated along such lines as social class, ethnic divisions, occupation and culture and these differentiated groups may not share the same rules. A very important consequence of this is that the person being judged as a deviant may feel that he is being condemned according to rules that he has had no hand in making and further that those whose social position gives them power and weapons are best able to enforce *their* rules. People who have the enterprise to undertake the enforcement of rules are called by Becker (1963) 'moral entrepreneurs'. They are represented by the crusader and they are dominated by those in the upper levels of the social structure.

A very controversial field is the way in which deviance affects society. One school of thought considers it as a kind of product of mental disease, others use models based on medical notions of health and disease and ask whether there are many processes going on in a society which reduce its stability and hence its survival chances. These processes would be considered as deviant and identified as symptoms of social disorganization. The most discussed theory put forward to explain deviance is probably Merton's theory of anomie. This implies that where there is a discrepancy between the goals of a society (e.g., 'The American Dream') and the chances of achieving those goals, there may be considerable deviant behaviour (crime, drug addiction, etc.). Some difficulties which arise in Merton's theory concern the problem of defining goals (functions) for a society which has many groups with many different rules. Lemert objects to Merton (in Clinard, 1964) on the grounds that the individual is not a free agent in his choice of values but rather is restricted by the claims of the various groups to which he belongs and that the nature of social value has not been sufficiently clarified. It has also been suggested that once an individual has been labelled as a deviant society tends to make him one, like the man in a German story who, having been in prison for some crime and having tried afterwards to get an honest living, was finally reduced to committing another crime to get back into prison (Fallada, 1934).

It is accepted by such sociologists as Clinard that there may be a definite connection between urbanism and deviance. The extensive conflict of norms found in urban industrial societies, the mobility both required and sometimes expected of the individual which leads to stresses and strains and the emphasis on individuation may lead, in certain urban areas, to deviance, though we cannot apply this to all urban districts. Some cities show less evidence of this kind of stress than others, but certain general cultural values in a society will actually increase the effects of urbanization, such as the emphasis on material possessions, or impersonality in human relationships which, in urban life, tends to foster individual freedom while at the same time creating what Riesman calls an 'inner directed' person (Riesman, 1950). There may be a decline in intimate communication:

> This stupid world where
> Gadgets are gods and we go on talking,
> Many about much, but remain alone,
> Alive but alone, belonging—where?
> Unattached as tumbleweed.
>
> (W. H. Auden, *The Age of Anxiety*, Random House, New York, 1946, p. 64.)

Suggestions for further reading

Books issued in paperback or cheap student editions are indicated by (p).

There is no general, comprehensive textbook of comparative urban sociology. The most useful introduction is provided by R. E. Pahl, *Readings in Urban Sociology* (1968p). This gives a survey of literature and topics in urban sociology and a number of extracts from literature published in Britain, the United States, Europe, Japan and Africa.

Other useful introductions are R. N. Morris, *Urban Sociology* (1968p), which gives an extensive analysis of the work of L. Wirth, and P. H. Mann, *An Approach to Urban Sociology* (1965p). In the USA there is A. Boskoff, *The Sociology of Urban Regions* (2nd ed., 1970).

For the strong-minded, the next step could be to tackle, in a selective manner, one of the massive 'Readers' which are produced in the USA. These usually contain material from other disciplines as well as urban sociology such as geography, economics and town planning, but it is not too difficult to find the relevant chapters.

Suitable readers are:

Hauser, P. M. and Schnore, L. F. (eds), *The Study of Urbanisation* (1965), especially chapters 5, 7, 10, 13, 14.

Hatt, P. K. and Reiss, A. J. (eds), *Cities and Society* (1957). The first chapter is valuable containing, as it does, a reprint of Wirth's essay, 'Urbanism as a Way of Life'. There is also a reprint of Simmel's article, 'The Metropolis and Mental Life', in chapter 8.

Burgess, E. W. and Bogue, D. J. (eds.), *Contributions to Urban Sociology* (1964). This selection mainly deals with material in the United States and some of the articles are reprints of older material.

Breese, G. (ed.), *The City in Newly Developing Countries* (1969) gives valuable extracts from urban sociological work in a variety of countries.

History and sociology

The initial reading should be Schnore's article in H. J. Dyos (ed.), *The Study of Urban History* (1968). This also contains Foster's study of class in certain Victorian towns. There is now a manageable selection from Charles Booth's *Life and Labour of the People in London: Charles Booth's London* (1971p), edited by A. Fried and R. Elman. E. Howard's book, *Garden Cities of Tomorrow*, has been reissued in a cheap edition (p).

G. Sjoberg, *The Preindustrial City* (1960p) and M. Weber, *The City* (1958p) are readily available. The latter book has a valuable introduction by Martindale.

Urban ecology: social structure and spatial structure

The best preliminary reading is to be found in R. E. Pahl, *Patterns of Urban Life* (1970). Though this book applies mainly to Britain, chapters III and IV discuss in outline the work of the Chicago school and on p. 37 there is a replica of Burgess's diagram. J. Madge, *The Origins of Scientific Sociology* (1963), has a chapter about the work of the Chicago school. The University of Chicago has reprinted much of its early work including studies by R. Park, and L. Wirth, *On Cities and Social Life* (1964p). Wirth is discussed critically by Morris (op. cit.), by L. Reissman, *The Urban Process* (1964) and in Pahl, *Readings in Urban Sociology*. For urban social structure, essential reading is J. M. Beshers, *Urban Social Structure* (1963) and Gluckman's essay in A. Southall (ed.), *Social Change in Modern Africa* (1961).

The family and urban living: networks

The most useful anthology is that of C. C. Harris—*Readings in Kinship in Urban Society* (1970)—which summarizes research in several fields. This leads on to C. Rosser and C. Harris, *The Family and Social Change* (1965), which can be compared with M. Young and P. Willmott, *Family and Kinship in East London* (1957p), and P. Willmott and M. Young, *Family and Class in a London Suburb* (1961p). For the study of network, necessary reading includes E. Bott, *Family and Social Network* (1957), and J. Mitchell (ed.), *Social Networks in Urban Situations* (1969).

Social divisions in urban living

Watson develops his ideas about 'spiralism' in a chapter in M. Gluckman (ed.), *Closed Systems and Open Minds* (1964). Studies of mobility in urban industrial societies are found in S. Lipset and R. Bendix, *Social Mobility in Industrial Society* (1959) and D. V. Glass, *Social Mobility in Britain* (1954). J. H. Goldthorpe *et al.*, *The Affluent Worker* (1968p) should be consulted for his concluding summary. In the USA, B. M. Berger, *Working Class Suburb* (1960p) and W. M. Dobriner, *Class in Suburbia* (1963p) will supplement reading on class.

Organizations

Voluntary associations in African towns are studied in K. L. Little, *West African Urbanisation* (1965p). Their relationship to social class in England is examined in D. V.

Glass, op. cit. (article by Bottomore), in Robson, op. cit., and in M. Stacey, *Tradition and Change* (1960).

Neighbourhood: community

Mann (op. cit., chapter 6) has summarized information about the neighbourhood and the neighbourhood unit. M. Stein, *The Eclipse of Community* (1960p) should be read in conjunction with P. Worsley (ed.), *Introducing Sociology* (1970p), chapter 6. The research work of R. N. Morris and J. Mogey, *The Sociology of Housing* (1965) and also of N. Elias and J. L. Scotson, *The Established and the Outsiders* (1965), have important implications for the study of communities.

The newcomers

L. F. Schnore and H. Fagin, *Urban Research and Policy Planning* (1967), have a chapter on 'acculturation' and 'assimilation' of immigrants, and Pahl (*Patterns of Urban Life*) discusses 'encapsulation' in Great Britain. J. Rex *et al.*, *Race, Community and Conflict* (1967p) has an important section on the Sociology of the Zone in transition, which should be read alongside the chapter by Gans in Pahl, *Readings in Urban Sociology*. O. Lewis's theory of urban poverty is to be found in the introduction to his book *La Vida* (1968p). Deviance is introduced by H. S. Becker, *Outsiders* (1963p) and is treated very fully in M. B. Clinard, *The Sociology of Deviant Behaviour* (3rd ed., 1968).

Bibliography

ADAMS, R. (1966), *Evolution of Urban Society*, Aldine Press, Chicago.

ALDOUS, J. and STRAUSS, M. (1966), 'Social Networks and Conjugal Roles, a test of Bott's hypothesis', *Journal of Social Forces*, no. 44, pp. 576-80.

ANDERSON, N. (1960) *The Urban Community*, Routledge & Kegan Paul.

BANTON, M. (1966), *The Social Anthropology of Complex Societies*, Tavistock.

BARCLAY, H. B. (1964), *Buuri al Lamaab*, Cornell University Press, New York.

BECKER, H. S. (1963p), *Outsiders*, New York and London.

BELL, C. R. (1968), *Middle Class Families*, Routledge & Kegan Paul.

BELL, N. W. and VOGEL, E. F. (1960), *The Family*, Chicago and London.

BENDIX, R. (1960), *Max Weber, An Intellectual Portrait*, Doubleday, New York.

BERGER, B. M. (1960p), *Working Class Suburb*, University of California Press.

BESHERS, J. M. (1963), *Urban Social Structure*, New York and London.

BEYER, H. G. (1967), *The Urban Explosion in Latin America*, Cornell University Press.

BLOOMBERG, W. and SCHMANDT, H. (1968), *Power, Poverty and Urban Policy*, Sage, New York.

BOOTH, C. (1902), *Life and Labour of the People in London*, Macmillan.

BOOTH, W. (1890), *In Darkest England and the Way Out*, Salvation Army International Headquarters.

BOSKOFF, A. (2nd ed., 1970), *The Sociology of Urban Regions*, Appleton-Century-Crofts, New York.

BOSKOFF, A. and CAHNMAN, W. T. (eds), (1964), *Sociology and History*, New York and London.

BOTT, E. (1957, 1971p), *Family and Social Network*, Tavistock.

BOTTOMORE, T. (1962p), *Sociology*, Unwin.

BREESE, G. (ed.) (1969), *The City in Newly Developing Countries*, Prentice-Hall, New York.

BURGESS, E. W. and BOGUE, D. J. (1964), *Contributions to Urban Sociology*, University of Chicago Press.

CHALMERS, T. (1821-6), *The Christian and Civic Economy of Large Towns*, Chalmers & Collins.

CHOMBART DE LAUWE, P. (1956), *La Vie quotidienne des familles ouvrières*, University of Paris thesis.

CHOMBART DE LAUWE, P. (1959-60), *Famille et habitation*, Centre Nationale de la Recherche Scientifique, Paris.

CHOMBART DE LAUWE, P. (1965), *Essais de sociologie 1952-1964*, Presses Universitaires de France, Paris.

CLINARD, M. B. (ed.) (1964), *Anomie and Deviant Behaviour*, New York and London.

CLINARD, M. B. (3rd ed., 1968), *The Sociology of Deviant Behaviour*, Holt, Rinehart & Winston, New York.

COATES, K. and SILBURN, R. (1970p), *Poverty; the Forgotten Englishmen*, Penguin.

COLLISON, P. (1963), *The Cuttleslowe Walls*, Faber.

Colloque de l'Ecole Normale Superieure de St Cloud (1967), *L'Histoire sociale: Sources et méthodes*, Presses Universitaires de France, Paris.

COTGROVE, S. (1967). *The Science of Society*, Allen & Unwin.

COULANGES, F. de (1864), *La Cité antique*, Hachette, Paris.

DAHRENDORF, R. (1959), *Class and Class Conflict in Industrial Society*, Routledge & Kegan Paul.

DAHRENDORF, R. (1965), *Gesellschaft und Democratie in Deutschland*, R. Piper Verlag, Munich; English trans. (1968), *Society and Democracy in Germany*, Weidenfeld & Nicolson.

DAVIS, K. (1955), 'The origin and growth of urbanisation', *American Sociological Review*, vol. 60, pp. 429-37.

DE VRIES, E. (1961), *Man in Rapid Social Change*, S.C.M. Press.

DOBRINER, W. M. (1963p), *Class in Suburbia*, Prentice-Hall, New York.

DORE, R. P. (1958), *City Life in Japan*, Routledge & Kegan Paul.

DUNCAN, O. D. and REISS, A. J. (1956), *Social Characteristics of Urban and Rural Communities*, Wiley, New York.

DUPONT-FERRIER, G. (1929), *La Formation de l'état français*, Librairie Armand Colin, Paris.

DYOS, H. J. (ed.), (1968), *The Study of Urban History*, Arnold.

EISENSTADT, S. N. (1970), *Readings in Social Evolution and Development*, Pergamon.

ELIAS, N., and SCOTSON, J. L. (1965), *The Established and the Outsiders*, Cass.

ENGELS, F. (1845, 1969p), *Condition of the Working Class in England*, Panther.

EPSTEIN, A. L. (1958), *Politics in an Urban African Community*, Manchester University Press.

EPSTEIN, A. L. (ed.), (1967), *The Craft of Social Anthropology*, Tavistock.

EPSTEIN, T. S. (1962), *Economic Development and Social Change in South India*, Humanities Press, New York.

ETZIONI, A. (ed.), (2nd ed., 1970), *A Sociological Reader in Complex Organisations*, Holt, Rinehart & Winston, New York.

FALLADA, H. (1934), *Wer einmal aus dem Blechnapf frisst*, Rowohlt Verlag Berlin English trans. (1970) E. Sutton, *Who Once Eats out of the Tin Bowl*, Howard Baker.

FAVA, S. (1968p), *Urbanism in World Perspective: A Reader*, Crowell, New York.

FESTINGER, L. A. *et al.* (1950), *Social Pressures in Informal Groups*, Tavistock.

FIRTH, R. *et al* (1969), *Families and their Relatives*, Routledge & Kegan Paul.

FRANKENBERG, R. (1966p), *Communities in Britain*, Penguin.

FREUND, J. (1968), *Sociology of Max Weber*, Allen Lane.

FRIED, A. and ELMAN, R. (ed.) (1971p), *Charles Booth's London*, Penguin.

FRÖHLICH, W. (ed.) (1964), *Afrika im Wandel seiner Gesellschaftsformen*, E. J. Brill, Leiden.

GANS, H. (1962), *The Urban Villagers*, New York and London.

GANS, H. (1967), *Levittowners*, Random House, New York.

GANS, H. (1968), *People and Plans*, Basic Books, New York.

GLASS, D. V. (1935), *The Town*, John Lane.

GLASS, D. V. (1954), *Social Mobility in Britain*, Routledge & Kegan Paul.

GLASS, R. *et al.* (1964), *London: Aspects of Change*, MacGibbon & Kee.

GLUCKMAN, M. (ed.) (1964), *Closed Systems and Open Minds*, Oliver.

GOLDTHORPE, J. H. *et al.* (1968-9p, 3 vols), *The Affluent Worker*, Cambridge University Press.

GOODE, W. J. (1963), *World Revolution and Family Patterns*, Collier-Macmillan.

HALMOS, P. (ed.) (1964), The Development of Industrial Societies, *Sociological Review Monograph 8*, University of Keele.

HALMOS, P. (ed.) (1966), Japanese Sociological Studies, *Sociological Review Monograph 10*, University of Keele.

HANDLIN, O. and BURCHARD, J. (eds) (1963p), *The Historian and the City*, Massachusetts Institute of Technology Press.

HARRIS, C. C. (1969p), *The Family*, Unwin.

HARRIS, C. C. (1970p), *Readings in Kinship in Urban Society*, Pergamon.
HATT, P. K. and REISS, A. J. (eds) (1957), *Cities and Society*, Chicago and London.
HAUSER, P. M. (ed.) (1957), *Urbanisation in Asia and the Far East*, UNESCO, Calcutta.
HAUSER, P .M. (ed.) (1961), *Urbanisation in Latin America*, UNESCO, Paris.
HAUSER, P. M. and SCHNORE, L. F. (1965), *The Study of Urbanisation*, Wiley.
HILLERY, G. A. (1955), *Communal Organizations*, University of Chicago Press.
HOWARD, E. (1946p), *Garden Cities of Tomorrow*, Faber.
JACKSON, J. A. (ed.) (1968), *Social Stratification*, Cambridge University Press.
JACKSON, J. A. (ed.) (1969), *Migration*, Cambridge University Press.
JANOWITZ, M. (1952), *The Community Press in an Urban Setting*, University of Chicago Press.
JANSEN, C. (1970p), *Readings in the Sociology of Migration*, Pergamon.
JONES, E. (1960), *Social Geography of Belfast*, Oxford University Press.
KHALDUN, IBN (1957), *The Muqaddinah*, Routlege & Kegan Paul.
KÖNIG, R. (1958), 'Gemeinde' and 'Gemeinschaft', pp. 73-88 in *Fischer-Lexicon 1958*, Fischer Bücherei, Frankfurt am Main.
KÖNIG, R. (1968), *The Community*, Routledge & Kegan Paul.
KUPER, L. (1953), *Living in Towns*, Cresset Press.
LEWIS, O. (1951p), *Life in a Mexican Village*, Holt, Rinehart & Winston, New York.
LEWIS, O. (1968p), *La Vida*, Panther.
LIPSET, S. M. and BENDIX, R. (1959), *Social Mobility in Industrial Society*, University of California Press.
LITTLE, K. L. (1965p), *West African Urbanisation*, Cambridge University Press.
LYND, R. S. and LYND, H. M. (1929), *Middletown*, Constable.
MABOGUNJE, A. L. (1968), *Urbanisation in Nigeria*, University of Liverpool Press.
MCGEE, T. G. (1967), *The South East Asian City*, Bell.
MCKELVEY, B. (1969p), *The City in American History*, Unwin
MADGE, J. (1963), *The Origins of Scientific Sociology*, Tavistock.
MANN, P. H. (1965p), *An Approach to Urban Sociology*, Routledge & Kegan Paul.
MARRIS, P. (1961), *Family and Social Change in an African City*, Routledge & Kegan Paul.

MARSH, C. (1965), *The Changing Social Structure of England and Wales*, Routledge & Kegan Paul.

MASTERMAN, C. F. G. (1909), *The Condition of England*, Methuen, chapter 3.

MAYER, P. (1961), *Townsmen or Tribesmen*, Oxford University Press.

MAYHEW, H., *London Life and the London Poor* (1861-5). A selection of texts is published by Spring Books and by the Oxford University Press.

MEADOWS, P. and MIZRUCHI, E. H. (1969), *Urbanism, Urbanisation and Change: Comparative Perspectives*, Addison-Wesley.

MEDHURST, D. F. and LEWIS, J. (1969), *Urban Decay*, Macmillan.

MERTON, R. K. (1949), *Social Theory and Social Structure*, Chicago and London.

MERTON, R. K. (ed.) (1959), *Sociology Today*, Basic Books, New York.

MERTON, R. K. *et al.* (eds) (1952p), *A Reader in Bureaucracy*, Chicago and London.

MICHELSON, W. H. (1970), *Man and his Urban Environment—a Sociological Approach*, Addison-Wesley.

MITCHELL, J. (ed.) (1969), *Social Networks in Urban Situations*, Manchester University Press.

MINER, H. (1939, 1963p), *St. Denis*, University of Chicago Press.

MOGEY, J. M. (1956), *Family and Neighbourhood*, Oxford University Press.

MORRIS, R. N. (1968p), *Urban Sociology*, Unwin.

MORRIS, R. N. and MOGEY, J. (1965), *The Sociology of Housing*, Routledge & Kegan Paul.

MORRIS, T. (1958), *The Criminal Area*, Routledge & Kegan Paul.

MOSER, C. A. and SCOTT, W. (1961), *British Towns*, Oliver.

MUMFORD, L. (1961p), *The City in History*, Penguin.

MUSGROVE, F. (1963), *The Migratory Elite*, Heinemann Educational.

OESER, O. A. and HAMMOND, S. B. (1954), *Social Structure and Personality in a City*, Routledge & Kegan Paul.

OSBORN, F. J. (1946), *Green Belt Cities*, Faber.

PAHL, R. E. (1968p), *Readings in Urban Sociology*, Pergamon.

PAHL, R. E. (1970p), *Patterns of Urban Life*, Longmans.

PARK, R. E. (1952), *Human Communities*, University of Chicago Press.

PARK, R. E. and BURGESS, E. N. (1925, 1967p), *The City*, University of Chicago Press.

PARKIN, D. (1969), *Neighbours and Nationals in an African City Ward*, Routledge & Kegan Paul.

PARSONS, T. (1937, 1968p), *The Structure of Social Action*, New York and London.
PARSONS, T. (1949, 1964p), *Essays in Sociological Theory*, New York and London.
PEACH, C. (1968), *West Indian Migration to Britain: a Social Geography*, Oxford University Press.
PERISTIANY, J. G. (1965), *Honour and Shame: The Values of Mediterranean Society*, Weidenfeld & Nicolson.
PFEIL, E. (1950), *Grosstadtforschung*, Walter Dorn Verlag, Bremen-Horn.
PHILLIPSON, J. (1966), *Ecological Energetics*, Arnold.
PIDDINGTON, R. (ed.) (1965), *Kinship and Geographical Mobility*, E. J. Brill, Leiden.
PIORO, Z. (1962), *Ekologia spoleczna w urbanistyce*, Arkady, Warsaw.
PIRENNE, H. (1956p), *Mediaeval Cities*, Princeton University Press.
PLOTNICOV, L. (1967), *Strangers to the City: Urban Man in Jos, Nigeria*, University of Pittsburgh Press.
PONS, V. (1969), *Stanleyville*, Oxford University Press.
REDFIELD, R. (1941p), *The Folk Culture of Yucatan*, University of Chicago Press.
REDFIELD, R. (1956p), *The Little Community*, University of Chicago Press.
REDFIELD, R. (1956p), *Peasant Society and Culture*, University of Chicago Press.
REISSMAN, L. (1964), *The Urban Process*, Collier-Macmillan.
REISSMAN, L. (1965p), 'Urbanism and Urbanisation', in *Social Sciences Survey*, Penguin.
REX, J. A., *et al.* (1967p), *Race, Community and Conflict*, Oxford University Press.
RIESMAN, D. (1950p), *The Lonely Crowd*, Doubleday, New York.
ROBSON, B. T. (1969), *Urban Analysis*, Cambridge University Press.
ROSS, A. D. (1961), *The Hindu Family in its Urban Setting*, Oxford University Press.
ROSSER, C. and HARRIS, C. (1965), *The Family and Social Change*, Routledge & Kegan Paul.
ROWNTREE, SEEBOHM B. (1901), *Poverty: A Study of Town Life*, Nelson.
RUEGG, W. (1969p), *Soziologie*, Fischer Bücherei, Frankfurt am Main.
SCHELSKY, H. (1965), *Auf der Suche nach Wirklichkeit*, Eugen Diedrich Verlag, Dusseldorf.
SCHNORE, L. F. (1965), *The Urban Scene*, New York and London.
SCHNORE, L. F. and FAGIN, H. (1967), *Urban Research and Policy Planning*, Sage, New York.

SCHORR, A. L. (1964), *Slums and Social Insecurity*, Nelson.
SCHULZ, D. A. (1969), *Coming up Black*, Prentice-Hall, New York.
SEELEY, J. R. *et al.* (1956), *Crestwood Heights*, Constable.
SHANIN, T. (ed.) (1971p), *Peasants and Peasant Societies*, Penguin.
SHEVKY, E. and BELL, W. (1955), *Social Area Analysis*, Stanford University Press.
SIMMEL, G. (1956), *The Sociology of George Simmel* (trans. K. Wolff), New York and London.
SJOBERG, G. (1960p), *The Preindustrial City*, Chicago and New York.
Social Implications of Industrialisation and Urbanisation in Africa south of the Sahara (1956), UNESCO, Paris.
SOUSTELLE, J. (1955p), *Daily Life of the Aztecs*, Penguin.
SOUTHALL, A. (ed.) (1961), *Social Change in Modern Africa*, Oxford University Press.
SOVANI, N. V. (1966), *Urbanization and Urban India*, Asia Publishing House.
SPROTT, W. J. H. (1958), *Human Groups*, Penguin.
SPYKMAN, N. J. (1966p), *The Social Theory of George Simmel*, Atherton, New York.
STACEY, M. (1960), *Tradition and Change*, Oxford University Press.
STEIN, M. (1960p), *The Eclipse of Community*, Harper & Row, New York.
SUTTLES, G. D. (1968), *The Social Order of the Slum*, University of Chicago Press.
THEODORSON, G. D. (1961), *Studies in Human Ecology*, Harper & Row, New York.
THOMLINSON, R. (1969), *Urban Structure*, Random House, New York.
THOMPSON, E. P. (1968p), *The Making of the English Working Class*, Penguin.
TIMMS, D. W. G. (1971), *The Urban Mosaic*, Cambridge U.P.
TOBY, J. (1964), *Contemporary Society*, Wiley.
TÖNNIES, E. (1955), *Community and Association*, Routledge & Kegan Paul.
TOURAINE, A. (1966), *La Conscience ouvriére*, Centre Nationale de la Recherche Scientifique, Editions du Seuil, Paris.
TOWNSEND, P. (1963p), *The Family Life of Old People*, Penguin.
TOWNSEND, P. *et al.* (1968), *Old People in Three Industrial Societies*, Routledge & Kegan Paul.
TURNER, R. (ed.) (1962), *India's Urban Future*, University of California Press.
UDRY, J. R. and HALL, M. (1965), 'Segregation and social networks in middle-class, middle-aged couples', *Journal of Marriage and the Family*, vol. 27, pp. 392-5.

VERNEY, S. (1969p), *People and Cities*, Fontana.
WAKEFORD, J. (1968p), *The Strategy of Social Enquiry*, Macmillan.
WARNER, W. L. (1941), *Yankee City*, Yale University Press.
WARNER, W. L. (1941), *Social Life of a Modern Community*, Yale University Press.
WARNER, W. L. (1949, 1960p), *Social Class in America*, Harper & Row.
WEBB, B. (1938, 1971p), *My Apprenticeship*, Penguin.
WEBER, A. F. (1899), *The Growth of Cities in the 19th Century*, Cornell University Press.
WEBER, M. (1947), *Theory of Social and Economic Organization* (trans. A. Henderson and T. Parsons), Collier-Macmillan.
WEBER, M. (1958p), *The City*, New York and London.
WHITTEN, N. E. (1965), *Class, Kinship and Power in an Ecuadorian Town*, Oxford University Press.
WHYTE, W. F. (1942, 1969p), *Street Corner Society*, University of Chicago Press.
WHYTE, W. H. (1957p), *The Organization Man*, Penguin.
WILLMOTT, P. (1963), *The Evolution of a Community; Dagenham*, Routledge & Kegan Paul.
WILLMOTT, P. and YOUNG, M. (1961p), *Family and Class in a London Suburb*, Routledge & Kegan Paul.
WILSON, G. and M. (1945, 1968p), *Analysis of Social Change*, Cambridge University Press.
WILSON, M. and MAFEJE, A. (1965), *Langa*, Oxford University Press.
WIRTH, L. (1928, 1964p), *The Ghetto*, University of Chicago Press.
WIRTH, L. (1964p), *On Cities and Social Life*, University of Chicago Press.
WORSLEY, P. (ed.) (1970p), *Introducing Sociology*, Penguin.
WURZBACHER, G. (1954), *Das Dorf im Spannungsfeld industrieller Entwicklung*, F. Enge Verlag, Stuttgart.
YOUNG, M. and WILLMOTT, P. (1957p), *Family and Kinship in East London*, Routledge & Kegan Paul.
ZORBAUGH, H. W. (1929), *The Gold Coast and the Slum*, University of Chicago Press.

For Product Safety Concerns and Information please contact our EU
representative GPSR@taylorandfrancis.com
Taylor & Francis Verlag GmbH, Kaufingerstraße 24, 80331 München, Germany

www.ingramcontent.com/pod-product-compliance
Ingram Content Group UK Ltd.
Pitfield, Milton Keynes, MK11 3LW, UK
UKHW022122190926
475702UK00013B/1588
* 9 7 8 0 4 1 5 8 6 4 6 8 8 *